T0327640

GROUNDING AND SHIELDING

GROUNDING AND SHIELDING

CIRCUITS AND INTERFERENCE

SIXTH EDITION

Ralph Morrison

For general information on our other products and services or for technical support, please
contact our Customer Care Department within the United States at (800) 762-2974, outside the
United States at (317) 572-3993 or fax (317) 572-4002.

Wiley also publishes its books in a variety of electronic formats. Some content that appears in
print may not be available in electronic formats. For more information about Wiley products, visit
our web site at www.wiley.com.

Library of Congress Cataloging-in-Publication data has been applied for

ISBN: 9781119183747

Typeset in 11/13pt TimesLTStd by SPi Global, Chennai, India

Printed in the United States of America

10 9 8 7 6 5 4 3 2 1

■■■ CONTENTS

Six editions of a book that spans 50 years is surely unique. I want to thank John Wiley for staying with me for all these decades. I want to thank Brett Kurzman, my new editor, for getting me a contract. I want to thank my wife Elizabeth for her continuous support. I want to thank all the readers that have supported me over the years. If it had not been for the urging of Dan Beeker of NXP Semiconductor, there would not have been a fifth edition. For this reason, I owe him a "thank you" for making a sixth edition possible. I have been puzzling how to make this edition more effective and I feel my opening statements are key. I need to tell the story so that the reader will appreciate my approach.

There are many meanings to the words grounding and shielding. To an English speaker, the nonengineering application of the word "ground" can include such diverse usage as coffee grounds, grounds for dismissal, playgrounds, ground round, or ground floor. The nonengineering use of shield can include such topics as windshield, police badge, metal armor, or protective clothing. In an electrical sense, ground can mean earth, the minus side of a battery, the conducting plane on a circuit board, the neutral power conductor, or a metal cabinet. People involved with electricity often associate these words with protection against electrical interference. The book title is intended to convey this meaning. Each reader of this book will start with a unique set of experiences associated with these words. I want to broaden that experience.

I have been involved in electrical grounding and shielding for over 50 years. My understanding comes from my experiences, my interests, and from my education. I have rewritten this book every 10 years since 1967 because the electrical world keeps changing. Also, I keep learning and the books do sell. Grounding and Shielding is an important topic as it relates to both cost and performance in about every aspect of our modern technology. The subject is difficult to present for many reasons.

First, it is related more closely to conductor geometry than it is to circuit content. Next, a lot of the information that is passed on through usage is simply lore and some of it is wrong or misleading. This means that often engineers have preconceived ideas and need to do some unlearning to get things straight. In some cases, the grounding rules a user must follow are a part of a code or a regulation that leaves no choices. When the rules are printed on fancy paper, it is easy to assume they are valid. If the rules are in error, arguing against the establishment can be very frustrating. Unfortunately, not all rules are effective or practical. And finally, the subject is not taught in schools and engineers are often on their own to find answers. I have also found that quality control people will follow written rules rather than the opinion of one outside engineer or author. They respect authority, which is what they are expected to do. I have had some polite arguments that have lasted years, where my viewpoint keeps being questioned. From this fact alone, I know just how ingrained some viewpoints can become.

The subject of grounding and shielding comes up in most designs. Because it is not an exact science, there can be many opinions as to where to connect shields and reference signals. Some approaches are lore and are passed on by copying past designs or by word of mouth. In most cases, there is no simple test that tells if a viewpoint is valid. We may know how to test a piece of hardware but testing a building is another matter. There is an important question that must be asked. What measurements would you like to make? Where do I put my voltage probes? Even if tests could be made, making changes to a large system can be very expensive. In many cases, the ideas used in a design are valid but over some limited range. A person with a misconception may find it difficult to accept a different explanation. This is especially true when many different opinions have been expressed. I find that for good reason, engineers are very skeptical about an explanation that suggests a new viewpoint. They do not know what or whom to believe. Another problem is that the engineering idiom is constantly changing and sometimes valid explanations fail to communicate. The subject lies somewhere between trade practices and physics and this is a wide separation. It is a subject area that does not fit into academia. To some managers, grounding seems like a job for a technician. In reality, it is often a very sophisticated and complex issue. For these and many other reasons, it is time to write a sixth edition. I want to try to get the message across for today's designers. As much as possible, I want to keep opinions to a minimum and I want to focus on connecting this subject to basic principles. I want to use physics as the basis for explanations but without getting too mathematical.

I have taken what I feel is a unique position in discussing grounding and shielding. Circuit theory leaves the distinct impression that conductors carry signals. The fact is that conductors guide the flow of field energy and this field energy can carry signals, interference, and operate components. Nature does not distinguish between these three functions. We definitely need circuits to describe our intentions. We need circuit theory to analyze these circuits. We need to apply basic physics when circuit theory is not sufficient. We have to keep power, signals, and interference separated because nature is not going to offer any help. We need to appreciate that nearly all electrical activity takes place in the spaces between conductors. It is the intent of this book to clearly point out how this very key idea can solve problems. This field transport idea is covered in physics courses. The connection between these ideas and practical designing is usually missing in textbooks and in the classroom. The arrangement of the spaces between conductors can keep various fields separated. This control of conductor geometry is the job of the designer. This is the central theme of this book.

This book is not an introduction to circuit theory. I assume the reader is familiar with how a circuit works. It is also not an introduction to applied physics. It is about all the things that happen when an education meets the real world. The problems that are solved in a textbook are nothing like the problems that are faced by the new designer. After graduation, there are no teachers around to explain how to proceed.

I solved many problems using my intuition and my past experiences. Little by little my understanding grew. Over time, it became clear that I should have used more of my education and less of my intuition. This is easy to say and not very easy to do. In the real world, problems must be resolved not studied. Most problems are multidimensional and do not fit just one subject. Finding a way through a maze may not be very efficient but experimenting with the maze to learn more is not often done. That takes time and that means spending resources.

I was told about displacement current in college. I never appreciated the value of this concept until I began writing about digital circuits. Then I discovered this idea let me explain how current flows into the distributed capacitance of a transmission line. In reading the literature, I had never seen this explanation. I present this idea in this book and I hope it helps to explain the real world to others.

In looking back at the fifth edition, I noticed how austere Chapter 1 appeared. Reading this material seemed a little like taking medicine to get well. It is not fun. Even though I wrote it with good intentions and I was excited to tell this story, the material did not appear inviting. To correct this condition, I decided to open this sixth edition with some of my background and describe a few of my experiences to illustrate why this basic physics is critical to an understanding of grounding and shielding. So stay with me as I go back in time. I think the history is both interesting and insightful.

July 1, 2015   RALPH MORRISON
San Bruno, CA

A HISTORICAL PERSPECTIVE INTO GROUNDING AND SHIELDING

As a teenager, in 1940, I built my own crystal set. I remember connecting the circuit common (ground) to a water pipe some 20 ft away from my set. The conductor went out through a hole in a wire mesh window screen. I found out that routing the antenna made a difference, so I kept trying different approaches to receive more radio stations. Little did I know of the complex nature of coupling to a transmitted radio signal. This was my first association with grounding (making an earth connection).

My interest in electronics extended to how a radio works and before long I was spending time in the neighborhood radio repair shop, where I learned how to test vacuum tubes. I was given a discarded radio as a present because the plastic case had been smashed. I freed the speaker cone and I had my own working radio. The radio had a ground clip that suggested that a grounding wire might improve reception. As a teenager, I spent time with several classmates that had acquired the skills needed to be amateur radio operators. They were always discussing their antennas and the grounding of their transmitters. I borrowed a copy of the ARRL handbook to get some idea of their hobby and what it meant to be a ham radio operator. I was an observer as I did not have the resources to enter this hobby.

I was drafted into the army in WW2 at age 18. Eventually, I became a radio repairman in the infantry and fixed radios in Patton's third army as it crossed Germany. The radios I serviced had no connections to earth as they had to be very mobile. I never gave grounding a second thought. After I returned home, the GI bill gave me the opportunity to go to Caltech and get a BS degree in physics. I remember taking courses in electricity and magnetism, not realizing the impact this subject would have on my future. I remember solving differential equations and fumbling through systems of units. I was introduced to Maxwell's equations. At the time, I had no way of assigning significance to this information. It was as if I was reading the first paragraph of many different chapters in many different books.

After graduation in 1949, I started working as an electronics engineer at a company called Applied Physics Corporation located in Pasadena, CA. My first boss was George W. Downs, a well-respected entrepreneur. During the war, he had worked as a high-level consultant and was associated with the Atomic Energy Commission. I had a lot to learn. The company products included oscillographs, electrometers, and spectrophotometers. I was impressed with the beautiful packaging and the fact they were so well respected by their customers. All of their products used vacuum tubes and I saw "grounding" for the first time. They explained to me how they used a grounding stud that collected all the common leads used in the instrument. This included the metal case, the equipment ground, the centertap on the secondary of the power transformer, the transformer shield, and the various circuit commons. There was no explanation given to me as to why this was the best solution. I was told that the order used in placing these conductors on the stud was important, and they had found a solution that made the instrument free of noise. In later years, this star-grounding configuration would appear in very unusual places. At the time, I had no basis to be critical of star-grounding methods. The products worked well and engineers with years of experience had spoken. Do not mistake me. A grounding stud was a valid approach to building this product. It is not however a solution to grounding in general. Asking questions did not yield useful answers and I did what everyone else did – I used common sense, I copied the procedures used in other products, and I experimented when I could. I was a part of the work force.

My first assignment as an engineer was to design a dc instrumentation amplifier. This type of instrument was needed in conditioning signals from strain gages, position sensors, and thermocouples. I was shown a circuit approach that had been developed by RCA that used a mechanical chopper to correct for dc drift. I was soon immersed in regulated dc power supplies, transformers, filaments, tube type selections, and feedback. I managed a design one channel of dc amplifier including a power supply that weighed over 70 lb. Do not forget that vacuum tubes take several hundred volts to operate and these voltages had to be very carefully regulated. When I look back at those early days, I can see how far electronic instrumentation has come and in particular how much I had to learn. At the beginning, there were no shielded transformers, feedback techniques were primitive, noise and hum were problems, and there was a limited understanding of signal isolation. There were selenium rectifiers that did not work very well. Dc amplifiers and vacuum tubes are a definite mismatch. In those days, that was all there was.

The techniques of differential amplification and common-mode rejection had not yet entered my understanding. My boss was learning from me. We had to start somewhere.

The period after WW2 saw the growth of the aerospace industry. I was project engineer on several analog computers that were sold to Douglas, Northrup, and Lockheed. These computers helped in the design of the first commercial jet aircraft. The computer design was based on work done at Caltech and included some dc amplifiers I had developed. After this project was completed, our instrumentation group was sold to a company in the transformer business. Our first project was to develop a high-speed recording oscillograph. The photographic paper speed in this machine was over 200 ft/s. Needless to say, Kodak appreciated our business. Getting the paper up to speed in milliseconds was no small task.[1] I designed the amplifiers that drove the galvanometers. I found out about the limitations imposed by using a common power supply to power a group of single-ended instruments. It was obvious that there was a lot to be gained by using a separate power supply for each signal channel. To meet this challenge, I began working on new techniques to reduce cost and size and avoid the use of common supplies. I invented a method of using AC coupling and a parallel feedback network to make a dc instrument. The company rejected my proposals for a new product line. I recognized the relevance of my new ideas, and I talked with two other engineers to leave and form a new company. George actually helped us make the transition.

The new company was called Dynamics Instrumentation. We manufactured instrumentation amplifiers for aerospace. The product line was based on the design ideas I had proposed. I could now contact users directly and I began to understand their dilemma. In rocket test stands, vacuum tube electronics had to be mounted in a blockhouse hundreds of feet away from the rocket engines and any sensors. This meant long input cables had to carry millivolt signals between structures. This raised issues of where to connect the input and output signal cable shields. I had some ideas on how to handle these issues and wrote some articles on the subject. I passed these articles out to potential customers. I was surprised at the reception these articles received. It was obvious there was very little information available on where to connect shields on large systems. Years later, engineers would pull these articles out of their files to show them to me. Now, when I look back at that period, I too had a

---

[1] The mechanical engineer was William McLellan also a Caltech graduate. He was the engineer that took the challenge from Dr. Richard Feynman to build a working motor 1/64 of an inch on a side. This motor is on exhibit at Caltech.

lot to learn. I could tell that this was a difficult problem and the size of a company had nothing to do with understanding the issues. Rocketdyne was interested in designing rocket engines not where to connect shields.

I found out that interference resulted from the flow of power current in input conductors. Remember there were hundreds of volts on the secondary coils of the power transformers. Simply put, I was the culprit. This current could be limited by the use of transformer shields. I built my own power transformers and played with the shielding until I understood what was happening. My competition built a carrier-type differential dc amplifier that used a mechanical modulator/demodulator and a multishielded input transformer. Being differential allowed input and output commons to be grounded separately without creating a ground loop. I tried to duplicate this approach, but I had problems building the input transformers. Instead, I built a postmodulator/demodulator around a postcarrier transformer using newly available transistors and managed in effect to build a wide-band differential amplifier. The mechanical modulator approach had 100 Hz bandwidth and the post-transistor modulator instrument I built had 10 kHz bandwidth. I had a new product and I had a new definition of the word differential as applied to instrument amplifier.

I needed three shields in the power transformers I used in this design. I got a company in San Diego to build them for me. I noticed one day in an electronics magazine that this company was offering what it called "isolation" transformers with four shields. On my next visit, I asked the company owner what his recommendation was for using a fourth shield. He did not know. I then asked why he offered it. The answer was simple: "They sell better." I had aided in the formation of a new business based on adding multiple shields to distribution transformers. I had used shields to make one instrument work and the industry had decided to use these same methods to "clean up" systems. To me, they had a solution looking for a problem. To me, the multishield solution only worked for one instrument. Later, I would take a broader view of this shielding. I also saw marketing at work.

I began to feel I had something important to offer other engineers. I saw a pattern in how shields worked and how they controlled the flow of interfering current. It was all tied to the electrostatics I had studied in school. So in the days of ribbon typewriters and carbon paper I started the task of writing a book. I showed the manuscript to Dr Ernst Guilleman of MIT and he was enthusiastic. I then submitted a manuscript to George Novotny the editor at John Wiley, where, much to my surprise, it was accepted for publication. The first edition of Grounding and Shielding was published in 1967.

My first analog designs used shielded conductors to carry signals on the circuit board. I was emulating the approach used at Applied Physics. When I designed my last instrument amplifier, there was not one shielded conductor inside the package. The circuit board did not have a ground plane. I had learned how to control the layout so that shielding signal leads was unnecessary. My approach was simple. I understood how to limit the area between conductors that carried the fields of the signals. I could run signals near power transformers and avoid even $1\,\mu V$ of coupling. The noise levels I attained were at the theoretical limits of the components and my instrument bandwidths were over $100\,kHz$. The users were still required to shield input signal cables between the transducers and the instruments.

Having a published book provided me with the opportunity to do consulting. I was now accepted by some as an expert. Selling instrumentation allowed me access to many military and aerospace facilities. This in turn allowed me to see how fairly large installations used their instrumentation. It was then that I began to see the conflicts imposed by regulations and controls. I found out that most of my recommendations were not followed. The engineers were happy because the performance was better than they had expected, and I was disappointed because things were just not very good. The engineers were constrained by rules that simply made no sense to me. As an example, I found that they had collected input and output shields in bundles and brought them to a single ground point. This was the star connection I had seen used at Applied Physics but applied in an entire building full of electronics. This is a good example of how lore can do damage when it is not controlled. In my early experiences, I expected logic to prevail and that I would be heard. I soon found out how much inertia there was and that the status quo would prevail. This was all a big disappointment. I was not heard. I was politely ignored. When I think about it, these engineers had little choice. The system was not designed to accept any step changes to approach. The bosses were from a different era and they wrote the rules. Somehow they got by and so did I.

I left Dynamics during an economic downturn and spent a few years in a company that built peripheral testing equipment for the telephone industry. The new building we were using had a so-called clean grounding conductor that was brought into the engineering area. The head of engineering had specified an approach that was a standard practice in telephone office design. Somehow the feeling that prevailed was that this was a "quiet" ground that would be used to reduce noise in testing hardware. I had a hard time accepting this strange logic, but I said nothing. This grounding rod was like the ones I had seen in aerospace.

It almost seemed that engineers had invented a new physics, where noise runs downhill into a sump and never returns. This violated my understanding that current flows in loops but here again these were experienced engineers doing the facility design. I wondered if the first user would contaminate the grounding rod for any subsequent users. During my years at this company, I never once used this "quiet" ground connection.

Before the era of fiber optics and microwave transmission, telephone links relied on copper connections. The fact that the entire country was crisscrossed with hard wire was impressive. To reduce costs, the ring (bell) circuits often used the earth as one of the conductors. This made it a necessary to provide a good earth connection at each central office. The head of engineering was simply following the good practice rules he had learned in the telephone industry.

At one point in my career, I needed some additional income. I noticed a company that offered a seminar called Grounding and Shielding. I contacted Don White Associates suggesting we might be able to work together as I had a book published with this same title. Don White agreed and I attended several of his grounding seminars. In this course, he focused his attention on the treatment of high-frequency interference. He discussed topics such as ESD, lightning, radar, rf filters, radiation standards, coaxial cables, and ground planes in computer installations. Even though I had been a physics major, I had a lot to learn. Eventually, I managed to catch up with the jargon and was able to teach courses in these new topics. I began to see that there was really no separation between analog and rf. It was one continuous subject. His course gave me an understanding of the specific problems engineers encountered in grounding and shielding at high frequencies. My analog design experiences did not include radiation. Don's experience did not involve instrumentation. Don gave me many opportunities to do consulting. Obviously, I had new material for another edition to my book. I also found out that the problems of radiation were far more common in industry than the problems I encountered in amplifying signals representing stress, strain, and temperature. Radiated interference that affected radio and television reception was regulated by the FCC. Interference that affected strain gage measurement was a minor distraction by comparison. There were no agencies regulating performance in instrumentation. I needed Don's help, but he did not need mine.

Living in Southern California gave me access to the aerospace industry, which included the Jet Propulsion Laboratories, aircraft companies, Edward Air force Base as well as Goldstone. I had two close friends,

Warren Lewis and Fred Kalbach that were full-time consultants. They often invited me to join them on their consulting field trips. Warren was invested in a business that sold power isolation transformers. One of these trips was to Goldstone, where lightning had knocked out a distribution transformer. This was potential business for Warren. Goldstone is where NASA located the antennas that communicated with deep space. It is located in the Mohave Desert far from habitation. This isolated location is necessary to avoid auto ignition noise as well as radio and television signals. The deep space antennas and the associated electronics are located in structures spread out over many acres of land. Just a reminder, these space antennas are nearly the size of football fields.

A single-point ground was provided for all signals and shields. This central grounding structure was a conducting rod placed in a deep well located at a point equidistant from each building. The power distribution transformers for each building were grounded locally per code. In this configuration, if lightning hit anywhere near the central grounding rod, the potential difference appearing across the windings in the distribution transformers could exceed their rated breakdown voltage. This indeed happened and they were blowing up transformers during summer thunderstorms. Here is a good example of where single-point signal grounding is not acceptable. The potential distributions in the earth that can occur during lightning activity are discussed later in the book.

The National Electric Code provides the rules used in distributing power to residences and facilities. The code was developed in the early history of power distribution under pressure from banks and insurance companies. Without controls, there were too many fires and lightning-related incidents. This code is constantly being revised to reflect best practice. In the mid-1980s, I wrote a book with Warren Lewis titled Grounding and Shielding in Facilities. I used the book as an opportunity to provide a rationale behind the code rules. In my consulting experience, I found engineers were often trying to avoid the code to solve some interference problem. If interference currents flowing in a neutral conductor were a problem, they pressured to unground the neutral. My job was to find another solution to the problem. An ungrounded neutral in a facility can be very dangerous. The code does permit this practice but only under carefully controlled conditions.

There was an economic downturn in 1991, and I decided to sell my second business. I had been involved in a small niche market. I had survived and earned a living, but I was never was able to grow as a company. After the sale, I spent my time writing and giving seminars. I received a call from Dan Beeker at Freescale[2] asking me if I could give a talk

---

[2] "Freescale is now NXP Semiconductor."

at a Freescale forum. He had found my name on one of my books in his library. That contact was the beginning of a long relationship that has lasted until the present. Freescale is in the semiconductor business and their customers design digital circuit boards. When the boards don't work, users blame the components. Freescale had to go on the offensive and help solve their problem or they would lose a customer. The reason is simple. It is easier to blame the semiconductors than to redesign the board. I worked with Dan and gave seminars to their customer base emphasizing transmission lines, the movement of energy, and the nature of interference. It took awhile, but Dan finally grasped the idea that the spaces on the board were more important than the traces. He became very effective in showing engineers how to avoid problems in board layout. As a result of this effort, I increased my understanding of key issues in multilayer printed circuit board design. I wrote a book titled *Digital Circuit Boards-Mach 1 GHz* that was published in 2012. Dan rarely has to design a board the second time. Layouts based on an understanding of fields do work. I have a lot to say in this book about energy flow on transmission lines. As logic gets faster and faster, a field understanding becomes a necessity.

Nature moves energy. She will take every opportunity to move to a lower potential energy state. If we are clever, we can use nature to do our bidding. We think in terms of voltage as we do not have energy probes. To be effective, we must understand how nature works. We must play by nature's rules. That is the subject of this book.

In this sixth edition I have rearranged the order of presentation to reflect current trends. I have added new material where appropriate, and I have dropped material that is no longer of interest. I hope this book will make it easier for engineers to do their job.

CHAPTER 1

Voltage and Capacitors

OVERVIEW

This first chapter describes the electric field that is basic to all electrical activity. The electric or E field represents forces between charges. The basic charge is the electron. When charges are placed on conductive surfaces, these forces move the charges to positions that store the least potential energy. This energy is stored in an electric field. The work required to move a unit of charge between two points in this field is the voltage between those two points.

Capacitors are conductor geometries used to store electric field energy. The ability to store energy is enhanced by using dielectrics. It is convenient to use two measures of the electric field. The field that is created by charges is called the D field and the field that results in forces is the E field. A changing D field represents a displacement current in space. This changing current has an associated magnetic field. This displacement current flows when charges are added or removed from the plates of a capacitor.

1.1 INTRODUCTION

Every person that has designed a circuit has considered issues of grounding and shielding. Every person that has used electronics to make measurements has encountered interference of some sort. The problems vary as the technology evolves. Searching for ways to deal with these issues consumes a lot of engineering time. It is the intent of this book to provide the reader with some insight as to what is happening. The sensing of signal takes place in the analog world. The computations involving data take place in the digital world. The energy to operate electronics comes from the power utility world. It makes

Grounding and Shielding: Circuits and Interference, Sixth Edition. Ralph Morrison.
© 2016 John Wiley & Sons, Inc. Published 2016 by John Wiley & Sons, Inc.

sense that the engineer must be familiar with all three areas if he wants to understand what is happening.

The parallel with automobile travel is interesting. A fine automobile makes sense only if there is an infrastructure. We need highways, bridges, repair centers, and gas stations or the system cannot work. The pieces of the system must work together for cars to be effective. Fuels must match engine needs, curves must match driving speeds, and the number of lanes must match traffic requirements.

This book makes an attempt to bring several disciplines together so that working in electronics is a lot easier. These disciplines involve the physics of electricity, the nature of the digital world, the shielding of the analog world, and finally the distribution of power in the utility world. To fully understand grounding, shielding, and interference, we must spend time in all of these areas. A discussion of utility power is important but unless transmission lines and radiation are understood, the subject of interference will make no sense. Shielding makes no sense unless the analog world is explained. There is a tendency to specialize in electronics. This book is an attempt to broaden the view and add to the general understanding of how nature functions.

How does a circuit work? One answer is to do a sinusoidal analysis using Kirchhoff's laws. Another answer is to write a set of logic statements. These responses provide a small part of the answer. The full answer is buried in a mountain of details. In this book, we are going to look at some of this detail but in a non-circuit way. We will take this approach because circuit diagrams and circuit theory by their very nature must leave out a lot of pertinent detail. This detail is important for quality performance whether it be for wide bandwidth or amplifying very low-level signals. It is also important when radiation, interference, or susceptibility is involved. Wire size, connection sequences, component orientation, and lead dress are often critical details. I like to call these details "circuit geometry." These details in geometry are important in analog circuits, power circuits, and especially in digital circuits, where clock rates rise year by year.

When a circuit is put to practice, there are many details that we take for granted. The components will most likely be connected together by strips or cylinders of copper. They will be soldered into eyelets or onto copper- or gold-plated pads. Traces will go between layers on a printed circuit board using vias. These are a few of the details in a design that are not questioned. There are details of a more subtle nature such as the thickness of a trace or ground plane or the dielectric constant of an epoxy board. In most cases, we do not question how things are done because we tend to rely on "accepted practice." Circuits built this way in the past have worked, so why make changes?

Taking things for granted is not always good engineering. Note that digital clock rates have changed from 1 MHz to 1 GHz in 20 years. That is three orders of magnitude! Imagine what would happen if automobile speeds went up one order of magnitude. That is 600 mph or jet aircraft speed. Even a modest increase in automobile speed would require extensive changes to the design of our roads and cities not to mention extensive driver training.

In electronics, an increase in speed does not pose a safety hazard. There are however differences and limitations in performance that should be understood. Often an effect is not sensed until the next generation in design is introduced. Understanding and correcting these effects requires an understanding of basic principles. The details we will look into do not appear on a circuit diagram. We will address these details because through understanding we can improve performance, reduce costs, and hopefully stay out of trouble.

Electronics often makes use of power from the local utility. For reasons of safety, the utility connects one of its power conductors to earth. Electronic hardware must often interface with this power and share this same earth connection. The result can be interference. I will discuss the relation between power distribution and circuit performance throughout the book.

A circuit diagram is only a plan or an organization of ideas. Circuit theory applied to a circuit diagram provides a basic overview of circuit performance. Circuit symbols are a part of the problem. They are necessarily very simple representations of complex objects. Every capacitor has a series resistance and an inductance. It can also be considered a transmission line stub. At some level it has its nonlinearities. Every inductor has series resistance and shunt capacitance. These considerations only begin to tell the entire story. For example, at high frequencies, dielectrics are nonlinear. For magnetic materials, permeability falls off with frequency. Thus, circuit symbols can only convey limited information. Further, we do not have symbols for skin effect, transit time, radiation, or current flow patterns. A straight line on a diagram may actually be a very complex path in the actual circuit. In short, a schematic diagram provides little information on physical structure and this can limit our appreciation of what is actually happening. If we had all this detail, we would be overwhelmed. It is important to be able to back off and take the broad view and still be aware that many details have been intentionally disregarded. A good designer walks a fine line, always aware that there are details in his field that he is not yet familiar with.

The performance of many circuits or systems is closely related to how they are built. It is not a question of whether there is an electrical connection but where the connection is to be made. In an analog circuit, it is often important to know which end of a shield is grounded not whether it is grounded. Here is a good question. How should a digital circuit board ground plane be connected to the surrounding chassis? The answer to this question is not available from a schematic diagram. Here is the answer: If possible, the connection should be arranged so that ground current does not flow in the ground plane of the board.

A repeated theme discussed in this book relates to how signals and power are transported in circuits. This approach will lead to an understanding of many issues that are often poorly understood. In order to discuss the transport of electrical power and signals, the electric and magnetic fields related to voltages and currents must be discussed. To begin this discussion, we introduce the electron. Don't despair. The time spent reviewing basic physics will make it much easier to understand the ideas presented in this book. Even if you do not follow the mathematics, the ideas will be clearly stated.

1.2 CHARGES AND ELECTRONS

Circuit theory allows us to relate circuit voltages and the flow of current in a group of interconnected components. For RLC networks (resistor–inductor–capacitor), this analysis is straightforward using Kirchhoff's laws. The processes I want to discuss do not involve this approach. To understand the fundamentals of circuit performance, we will use basic physics to explain many details that are often ignored. Our starting point may seem a bit remote but please read on.

Atoms are composed of a nucleus of protons and neutrons surrounded by electrons. For our purposes, the electrons can be considered negatively charged particles located in shells around the atom. The quantum mechanical view of electrons in atoms is that they are overlapping waves described by quantum numbers and a probability function. We will use the shell viewpoint of the atom as we are not involved in nuclear physics or quantum mechanics. We can treat electrons as particles and not get into trouble. The electrons have a negative charge and the matching protons in the nucleus have a positive charge. In a neutral atom, the positive and negative charges are exactly equal. Each electronic shell is limited to a fixed number of electrons. The number of electrons in the outer shell says a lot about the character of the atom. As an example, copper has just one electron in its outer shell. This outer electron has a great deal of mobility and is involved in electrical conductivity. Because protons are comparatively heavy and the shells of electrons shield them, they are not directly involved in the electronics we are going to consider.

Molecules are formed from atoms that bond together. Bonding really means that electrons from one atom share spaces in the shells of other atoms. For an insulator, this bonding greatly restricts the activity of outer shell electrons. Typical insulators might be nylon, air, epoxy, or glass. This bond does vary between insulators. If two insulators are rubbed together, such as a silk cloth against a rubber wand, some of the shared electrons on the wand will transfer to the cloth. In this case, the silk cloth with extra electrons is called a negatively charged body. We will call the absence of negative charge a positive charge. The rod is said to be positively charged. In reality, the positive charge stems from the immobile protons in the nucleus of atoms that do not have matching outer shell electrons. The absence of negative charges behaves the same as if there were fictitious positive charges on the surface of the insulator. An analogy can help. Consider an auditorium full of people. There is one empty seat. When a person moves left into this seat, the vacant seat has moved to the right. If people keep changing seats this way, we can see the flow as moving empty seats. Now consider the empty seat as a positive charge.

Experiments with charged bodies can demonstrate the nature of the forces that exist between charges (electrons). These same forces exist for the absence of charge that we will call a positive charge. If one charged body repels another, it is actually the fields of electrons that are involved. If you remember your physics class, these forces can be demonstrated using pith balls that hang by a string. Here the charges are attached to small masses and we can see the pith balls attract or repel each other. For there to be an attraction, one pith ball needs to have extra electrons and the other an absence of electrons.

The percentage of electrons involved in any of these experiments is extremely small. To illustrate this point, I want to paraphrase the writing of Dr. Richard Feynman.[1] If two people are standing a few feet apart, what would be the force of repulsion if 1% of the electrons in each body were to repel each other? Would it be a few pounds? More! Would it be greater than their weight? More! Would it lift a building? More! Would it lift a mountain? The answer is astounding. The force would be great enough to lift the earth out of orbit. This is

[1] The Feynman Lectures on Physics Volume 2 page 1-1. Addison Wesley Publishing Company Inc. Copyright 1964, California Institute of Technology.

why gravity is called a weak force and the force between electrons is called a strong force. This also tells us something about nature. The percentage of electrons involved in electrical activity is extremely small. We know that the forces in a circuit do not move the components or the traces. Obviously, since electrical forces are so large, electrical activity in a circuit involves an extremely small percentage of the available electrons. Yet there are so many electrons on the move, we can think of any typical current flow as being continuous and seamless.

1.3 THE ELECTRIC FORCE FIELD

When we encounter forces at a distance, we use the expression force field. We experience a force field at all times as we live in the gravitational force field of the earth. Every mass has a force field including the earth. The earth has the dominant field because the earth is so massive. The result is that each mass on earth is attracted toward the center of the earth. The forces of attraction between individual objects on the earth are so small that they are very difficult to measure. On the earth's surface, the force field is nearly constant. We would have to go out thousands of miles into space to see a significant reduction in the force of gravity.

The electrical and gravitational force fields are similar in many ways. Every electron carries with it an associated force field. This force field repels every other electron in the area. If a group of extra electrons are located on an isolated mass, we call this mass a charged body. We refer to the extra electrons as a charge. If this mass is a conductor, the extra electrons will move apart until there is a balance of forces. On a conducting isolated sphere, the extra electrons will move until they are evenly spaced over the entire outer surface. None of these excess electrons will remain on the inside of the conductor. For a perfect insulator, extra electrons are not free to move about. Extra electrons on the inside of this material are called trapped electrons. It is also possible to have trapped absences of electrons.

1.4 FIELD REPRESENTATIONS

The electric force field in a volume of space can be measured by noting the forces on a small test charge in that space. A test charge can be formed using a small mass with a small excess of electrons on its surface. The force on this test charge has a magnitude and direction at each

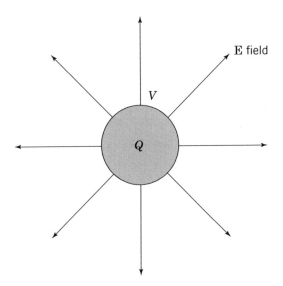

Figure 1.1   The force field lines around a positively charged conducting sphere

point in space. Having direction, the force field is called a vector field. To be effective, this test charge must be small enough so that it does not influence the charge distribution on the objects being measured. Performing this experiment is difficult but fortunately we can deduce the field pattern without performing an actual test.

It is convenient to represent a force field by lines that follow the direction of the force. For an isolated conducting charged sphere, the lines of force are shown in Figure 1.1.

Note that the field exists everywhere between the lines. The lines are simply a way of showing the flow or shape of the field. As pointed out, the number of extra electrons that form the charge Q on the surface of a conductor is small compared to the number of electrons in the conductor. In spite of this fact, the number of electrons is still so large that we can consider the charge as being continuously distributed over the surface of interest. This is the reason we will not consider the force field as resulting from individual electrons. From here on, we will consider all charge distributions as being continuous. The total charge on the surface of the sphere in Figure 1.1 is Q. The charge density on the surface of the sphere is

$$\frac{Q}{A} = \frac{Q}{4\pi r^2}. \tag{1.1}$$

We will use the convention that a line starts on a unit of positive charge and terminates on a unit of negative charge. This unit can be selected so that the graphical representation of the field is useful. If the total charge is doubled, then the number of lines is doubled. For representations in this book, no attempt will be made to relate the number of lines to any specific amount of charge. In general, we are interested in the shape of the field, areas of field concentration, and where the field lines terminate. For the ideal single sphere, the field lines are straight and extend from the surface to infinity.

In Figure 1.1, the force f on a small test charge q in the field of a charge Q located on a sphere is proportional to the product of the two charges and inverse to the square of radius r or

$$f = \frac{qQ}{4\pi\varepsilon_0 r^2} \tag{1.2}$$

The constant ε_0 is called the permittivity of free space. Equation 1.2 is known as Coulomb's law. The force per unit charge or f/q is a measure of the electric field intensity. The letter E is used for this measure. The force field around a group of charges is referred to as an E field. Mathematically, the E field around a charge Q is

$$E = \frac{Q}{4\pi\varepsilon_0 r^2} \tag{1.3}$$

The E field falls off as the square of the distance r. In Figure 1.1, the force field intensity E decreases as the field lines diverge. The forces are greatest at the surface of the sphere. Note that the field lines do not enter the sphere. This is because there are no excess charges inside the conductor. The field lines must terminate on the sphere perpendicular to its surface. If there were a tangential component of force on the surface, the charges on the surface would be accelerated. If there were an absence of electrons on the surface, this absence of charge would also be accelerated. Remember the absence of negative charge can be considered the presence of a positive charge. For conductors, the mobility of a group of electrons is no different than the mobility of an absence of electrons. Except for the direction assigned to the force field, we will assume that positive and negative charges behave the same way. Figure 1.1 shows a sphere with a positive charge Q. If the charge were negative (the presence of electrons), the field lines would be shown with the arrows pointing inward.

The field lines in Figure 1.1 start at the surface of the sphere. If the charge Q were located at the center of the sphere and the sphere were removed, the field pattern at every initial value of r would be unchanged. A point charge Q implies an infinite charge density, which is impossible. Often it is mathematically convenient to consider the fields from point charges even though this cannot exist.

N.B.

The electric force field E is called a vector field as it has a magnitude and direction at every point in space.

N.B.

The field intensity is greatest where the lines are closest together.

1.5 THE DEFINITION OF VOLTAGE

A test charge q in the field of a charge Q experiences a force f given by Equation 1.2. The work required to move the test charge a small distance Δd is $f \cdot (\Delta d)$. The work to move it from infinity to a point r_1 is the integral of force times distance from infinity to r_1. If we follow one of the field lines, the force is always tangent to this line. The work is

$$W = \int_{\infty}^{r_1} f \ dr = -\frac{qQ}{4\pi\varepsilon_0 r_1}.$$ (1.4)

If we divide both sides of this equation by q, we obtain the work per unit charge. This term has the familiar name volts. In equation form, the voltage V is given by

$$V = \frac{Q}{4\pi\varepsilon_0 r}.$$ (1.5)

DEFINITION

A voltage difference is the work required to move a unit charge between two points in space in an electric field.

In Equation 1.4, we can make the assumption that the voltage at infinity is zero. This allows us to assign a voltage to intermediate points in space. In a circuit, the work required to move a unit charge between two conducting surfaces is called a potential difference or a voltage difference. It is important to realize that potential differences do exist between points in space. Of course, it is difficult to place a voltmeter in space to get a measure of this voltage.

The voltage difference between two points in space is

$$V_2 - V_1 = \left(\frac{Q}{4\pi\varepsilon_0}\right)\left(\frac{1}{r_2} - \frac{1}{r_1}\right).$$ (1.6)

N.B.

A voltage difference cannot exist without the presence of an electric field.

In the presence of conductors, an electric field cannot exist without charges on the surface of these conductors. These charges are not apparent from a schematic diagram.

When a circuit is in operation, there are surface charges everywhere and there are voltage differences. These surface charges are the first charges that move when a request is made to move energy. In a dc circuit, electrons move in the entire conductor. A current is considered dc only after a period of time has elapsed and no further change in the flow pattern can be detected.

N.B.

Fields and charges are not shown on schematic diagrams.

1.6 EQUIPOTENTIAL SURFACES

As the word implies, an equipotential surface is a surface of equal voltage. No work is required to move a test charge on this surface. This surface can be in space or on a conductor. Figure 1.2 shows equipotential surfaces around the charged sphere in Figure 1.1. Note that these surfaces are also spheres and the E field lines are always perpendicular to these equipotential surfaces.

N.B.

Conducting surfaces are equipotential surfaces regardless of their shape. This assumes that the surface charges are not in motion.

In practice, a conducting surface is an equipotential surface even when the charge distribution is not uniform. This is true even when there are

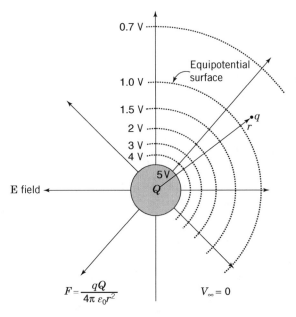

Q is the charge on the sphere
q is small unit test charge
ε_0 is the dielectric constant of free space (permittivity)

Figure 1.2   Equipotential surfaces around a charged sphere

areas of positive and negative charges on the same conductor. We will assume it takes no work to move charges on the surface of a conductor. If work were required, there would be a tangential electric field and this means that free charges would have to be in motion.

1.7 THE FORCE FIELD OR E FIELD BETWEEN TWO CONDUCTING PLATES

Consider two conducting plates separated by a distance h. On the top plate there is a charge $+Q$ and on the bottom plate there is a charge $-Q$. This configuration is shown in Figure 1.3.

If we ignore edge effects, the force field can be represented by equally spaced straight lines that run from the top plate to the bottom plate.[2] In this configuration, the net charge in the system is zero. There is no loss in generality if we assume that all of the field lines stay in the volume between the two plates. Since the lines do not diverge, the force on a test charge q is constant everywhere in the space between the plates. In other

---

[2] Edge effects can be ignored when the spacing between conductors is very small.

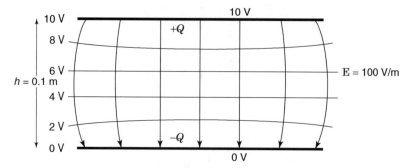

Figure 1.3   The force field between two conducting plates with equal and opposite charges and a spacing distance h

words, the electric field intensity E is a constant between the plates. If the charge density on the plates is made equal to the charge density of the sphere in Figure 1.1, the force field between the plates will have the same intensity as the force field at the surface of the sphere. The work required to move a unit charge between the plates will then be force times distance or

$$W = \frac{Qh}{4\pi\varepsilon_0 r^2} = Eh. \tag{1.7}$$

This work W is the potential difference between the plates. If the bottom plate is assumed to be at 0 V, the upper plate will be at a voltage V = Eh. Note that the E field has units of volts per meter. If the voltage between the two planes of Figure 1.2 is 5 V and the spacing is 1 cm, the E field intensity between the plates is 500 V/m.

1.8 ELECTRIC FIELD PATTERNS

Figure 1.4 shows a printed circuit trace over a conducting surface. Such a surface is often called a ground plane. There are many types of ground planes and we will talk about this later. The spacing between these two conductors might be as small as 0.005 in. or 1.3×10^{-4} m. A typical logic voltage might be 5 V. The E field intensity under the trace would be 38,000 V/m, a very surprising figure.

The E field lines terminate and concentrate on the surfaces between the trace and the ground plane. Field lines terminate on charges. Remember this is a static situation. Note that surface charge distributions are not considered in circuit theory. The path taken by the charges to achieve this distribution is also not considered.

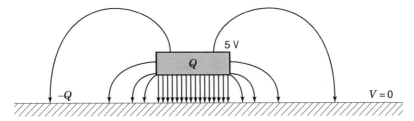

Figure 1.4   The electric field pattern of a circuit trace over a ground plane

---

FACTS

1. The charge distribution on the surface of conductors is usually not uniform.
2. There is no potential gradient along the surface of the ground plane as the charges are not moving.
3. The charges concentrate at the surfaces between the circuit trace and the ground plane.
4. There is some E field above the circuit trace.
5. Surface charges concentrate on the sharp edges of the circuit trace.
6. If the voltage were to increase slowly, new charges must be supplied to the conductors.

Consider the field pattern when there are two traces over a ground plane. This pattern is shown in Figure 1.5. If the voltages are of opposite polarity, the charge distribution on the ground plane will reverse

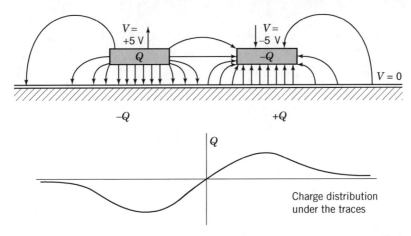

Figure 1.5   The electric field pattern around two traces over a ground plane

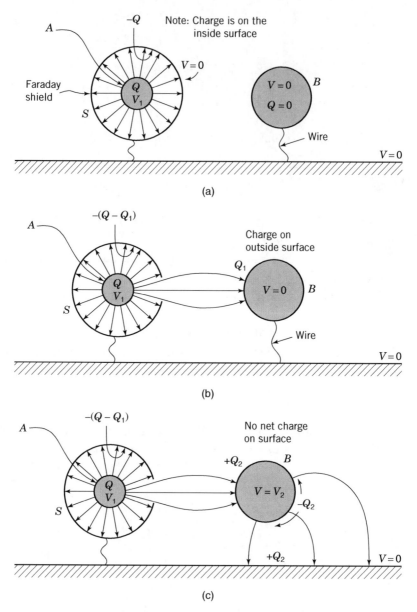

Figure 1.6   Field configurations around a shielded conductor

polarity under the traces. Again, the ground plane is an equipotential surface.

Consider the field pattern around a section of shielded cable as in Figure 1.6a.

In Figure 1.6a, the shield S fully encloses the center conductor A and no electric field escapes. In Figure 1.6b and c, there is a hole in the shield that allows some of the electric field to terminate on conductor B. The field lines that terminate on conductor B imply a charge distribution on conductor B. In Figure 1.6c, conductor B floats in space. Note there is still a charge distribution on this conductor, but the net charge on the conductor is zero. This floating conductor would not be at zero potential, but it would still be an equipotential surface. The grounded conductor by definition would be at zero potential over its entire surface even though there is a net charge on its surface.

If the voltage on the shielded conductor in Figure 1.6b is slowly changed, the field intensity changes everywhere. The amount of charge on conductor B in Figure 1.6b must now change. This change in charge must flow in the connection to the ground plane. The charge on conductor B is called an induced charge. The flow of charge is called a current. Any current that flows to conductor B is called an induced current. In the case of Figure 1.6c, the charge density on the surface of the floating conductor must change. This means that induced currents must flow locally on this conductor. Note there is no path for new charges to reach the isolated conductor. There simply needs to be a change in the electric field in the space around the conductor for there to be a change in the charge distribution.

N.B.

Surface currents can flow on a conductor that is floating, that is, not connected to any other circuit.

The electric field in Figure 1.6a is totally contained. The internal field can change and there are no induced currents on nearby conductors. This containment of the electric field is called electric shielding. The outer conductor S is often called a Faraday shield. A conductor with an outer conducting sheath is called a shielded cable. Later we will discuss a shielded conductor called coax.

N.B.

The electric field lines in Figure 1.6 terminate on the inside surface of the cable. If the voltage changes slowly, the resulting change in field causes current flow on the inside surface of the shield. Ideally, this field does not penetrate into this shield and get to the outside surface.

N.B.

Shielding has nothing to do with external connections to the shield conductor. If the electric field is contained, the shield is effective. The shield need not be at "ground potential" to be effective.

1.9 THE ENERGY STORED IN AN ELECTRIC FIELD

It takes work to move a charge in an electric field. In Figure 1.3, the work required to move a unit charge between the plates is the voltage difference between the plates. As more charge is moved across the space, the voltage between the plates increases. The work done on the charges is stored as potential energy. Where is this energy stored? Since it is not stored inside the conductors or on their surfaces, the only place that is left is in the space between the plates. The same problem exists with gravity. When a weight is lifted in a gravitational field, the added energy is stored in this field not in the mass.

The force on a small increment of charge dq is $\mathrm{E}\,dq$, where E is the force field. Assume the top plate has a charge q and the bottom plate has a charge $-q$. The E field from Equation 1.3 is $q/\varepsilon_0 A$.

The work dW required to move an increment of charge dq across the distance h is

$$dW = f\,h\,dq = \left(\frac{q}{\varepsilon_0 A}\right) h\,dq. \tag{1.8}$$

By integrating Equation 1.8 over charge from 0 to Q, the total work W required to move a charge Q is

$$W = \int_0^Q \frac{q\,dq}{\varepsilon_0 A} = \frac{Q^2 h}{2\varepsilon_0 A}. \tag{1.9}$$

Since $\mathrm{E} = Q/(\varepsilon_0 A)$, the work W can be written as

$$W = \frac{\mathrm{E}^2\,\varepsilon_0 A h}{2} = \frac{1}{2}\mathrm{E}^2 V \varepsilon_0 \tag{1.10}$$

where V is the volume of the space between the two conductors.

N.B.

Every volume of space that contains an E field stores electric field energy.

N.B.

Every circuit with voltages stores electric field energy.

Somehow space has a quality like a spring that can be used to store potential energy. The conductors seem to provide a handle for holding on to this spring. We cannot see it or feel it, yet we can do work with the energy that is stored in space. Without the restraint of these conductors, the field energy would have to leave the area at the speed of light. The correct word here is radiation.

In practical circuits, the electric field patterns are complex and the intensity of the field varies over space. To calculate the total stored energy, space can be divided into small volumes of near-constant field intensity. The important fact to remember is that the energy stored per unit volume is proportional to the square of the field intensity. In many practical problems, the region of high field intensity is all that is important. Remember this is where most of the energy is stored.

The potential assigned to a point in space is a number not a vector. Some surface in the system must be assigned to the zero of potential. The potential at every other point in the system is the work required to move a test charge against the forces created by every element of charge in the system. This voltage value is a summation based on Equation 1.5. The electric field vector can be determined by locating the direction in which the voltage change is maximum. Stated mathematically, the gradient of the voltage is the electric field intensity.

The electric fields we will consider in circuits are located in the spaces between conductors. These fields terminate on charges distributed on the conducting surfaces. There can be no electric field in the conductors or there would be current flow. This means there is essentially no field energy stored in the conductors that is available to us to do work. This supports the theme of this book that the spaces between conductors carry the fields that represent signal, energy flow, as well as interference.

1.10 DIELECTRICS

We next consider the effect dielectric materials have on electric fields. Typical dielectrics are rubber, silk, Mylar®, polycarbonate, BST, epoxy, air, and nylon. Up until now, we have considered the electric field in air (technically in a vacuum). Consider the two plates in Figure 1.3. If the space between the plates is filled by an insulating dielectric, it takes less work to move a charge Q from one plate to the other. This means the force field inside the insulator is reduced. This reduction factor ε_R is known as the relative dielectric constant. The force field between the planes in the dielectric medium is given by

$$E = \frac{Q}{A\varepsilon_0\varepsilon_R}. \qquad (1.11)$$

If the space is first filled with air, then a voltage V results in a charge Q. When a dielectric material is inserted between the plates, the voltage must drop to V/ε_R. If the voltage is again increased to V, the amount of charge on the surface would increase by the factor ε_R. This factor is called the relative dielectric constant. This factor is exactly 1 in free space.

N.B.

The relative dielectric constant of air is 1.0006.

N.B.

Dielectric materials are used in capacitors to increase the charge stored per unit voltage.

1.11 THE D FIELD

It is convenient to discuss two measures of the electric field. The voltage between two points defines the E field intensity. A second field measure called the D field relates directly to charges. In a vacuum, the E field and D field patterns are exactly the same. In a region where there are dielectrics, the E field intensity changes at every dielectric interface. The D field starts and stops on charges but does not change intensity at a charge-free boundary. In Figure 1.1, if the charge Q were located in a dielectric medium, the E field would be reduced by a factor equal to the relative dielectric constant. The new E field is then given by Equation 1.12.

$$E = \frac{Q}{4\pi\varepsilon_0\varepsilon_R r^2}. \qquad (1.12)$$

The energy stored in the field of this charge is inversely proportional to the relative dielectric constant. Figure 1.7 shows the field pattern

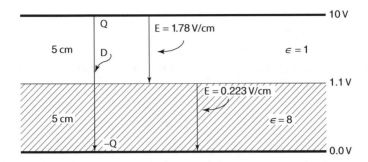

Figure 1.7   The electric field pattern in the presence of a dielectric

between two planes, where half of the space has a dielectric constant of 8. If the total spacing is 10 cm and the dielectric material has a dielectric constant 8, the E field pattern must adjust so the total voltage difference is 10 V. The voltages in terms of the E field are $(E/8) \cdot 5$ cm $+ (E/1) \cdot 5$ cm $= 10$ V. E in the open space is obviously 8.9 V/cm and inside the dielectric it is 1.1 V/cm. The voltages are 8.9 V across the air space and 1.1 V across the dielectric. The electric field intensity in air is now 8.9 V/5 cm $= 1.78$ V/cm. Before the dielectric was inserted, the field intensity in air was 1 V/cm. This means that the added dielectric has increased the charge Q on the plates by 78%. Note that the majority of energy is stored in the air space not in the dielectric. In Figure 1.7, the D field is continuous from the top to bottom plates. If E equals D/ε_0 in the air space, then in the dielectric

$$E = \frac{D}{\varepsilon_R \varepsilon_0}. \tag{1.13}$$

N.B.

Electric field energy is stored in the E field.

In high-voltage transformers, an oil dielectric is often used to reduce the E field around conductors. This reduction in the E field reduces the chance of arcing. The oil also helps to conduct heat away from the windings.

1.12 CAPACITANCE

The ratio of charge to voltage is capacitance C. The unit of capacitance is the farad F. A capacitor of 1 F stores one coulomb of charge Q for a voltage of 1 V. In Figure 1.1, the voltage on the surface of the sphere is associated with a stored charge Q. The voltage V is $Q/4\pi r\varepsilon_0$. The ratio Q/V equals $4\pi r\varepsilon_0$. If the sphere is located in a dielectric medium, the voltage V is reduced by ε_R and the ratio Q/V is $4\pi r\varepsilon_0\varepsilon_R$. The capacitance of a conducting sphere in a dielectric medium is

$$C = 4\pi r\varepsilon_R\varepsilon_0. \tag{1.14}$$

For the parallel planes in Figure 1.3, the voltage between the conducting plants is the E field times the spacing h. The voltage from Equation 1.11 is $V = Qh/(\varepsilon_0 A)$. If there is a dielectric present, the ratio Q/V is

$$C = \frac{\varepsilon_0\varepsilon_R A}{h}. \tag{1.15}$$

Capacitance is a function of conductor geometry. So far, we have discussed two simple geometries: the sphere and parallel conducting planes. In most practical circuits, the geometries are complex and the capacitances are not simply calculated. It is important to recognize that capacitance is controlled by three factors. It is proportional to surface area, inversely proportional to the spacing between surfaces, and proportional to the relative dielectric constant.

N.B.

Capacitance is a geometric concept. All conductor geometries can store some electric field energy; therefore, they all have capacitance.

The idea of capacitance can be extended into free space. Consider a cube in space oriented so that electric field lines are perpendicular to two of the cube's faces. If there were equal and opposite charge distributions on the two opposite faces of the cube, the field lines would be no different. The voltage between the faces is the E field times the distance across the cube. Since we have an equivalent surface charge and a potential difference, the ratio is capacitance.

N.B.

Free space has the ability to store electric field energy. A volume of space has a capacitance. An electric field cannot use this space to store static energy unless there are nearby conductors to hang on.

The factor ε_0 is called the permittivity of free space and it is equal to 8.85×10^{-12} F/m. Consider the capacitance of a printed circuit trace over a ground plane. If the spacing h is 5 mm and the trace is 10 mm wide by 10 cm long, the trace area is 100 mm². The value of A/h is 20 mm or 20×10^{-3} m. If the relative dielectric constant is 10, then the capacitance between the trace and the ground plane is equal to $(A/h)\varepsilon_R\varepsilon_0 = 177 \times 10^{-12}$ F or 177 pF.

It is interesting to calculate the capacitance of the earth as a conductor. The radius of the earth is 6.6×10^6 m. Using Equation 1.14, the capacitance is 711 μF.

1.13 MUTUAL CAPACITANCE

A mutual capacitance is often referred to as a leakage capacitance or parasitic capacitance. The electric field pattern in most practical circuits is complex. The voltage on any one conductor implies a self-charge and induced charges on all the other conductors. For small component geometries, a large percentage of the field energy may be parasitic in nature.

The ratio of charge to voltage on any one conductor is called a self-capacitance. Examples of self-capacitance are shown in Figures 1.1 and 1.4. The ratio of the charge induced on a second conductor to voltage on a first conductor is called a mutual capacitance. An example of a mutual capacitance is shown in Figure 1.6b. A measure of this capacitance requires a test voltage be placed on one conductor and all other conductors must be at zero potential. The mutual capacitance C_{12} is the ratio of charge induced on conductor 2 for a voltage on conductor 1. It turns out that $C_{12} = C_{21}$.[3]

All mutual capacitance values are negative as the induced charge for a positive voltage is always negative. A simple geometry showing a few mutual capacitances is shown in Figure 1.8.

A voltage V_1 is placed on trace 1 and traces 2, 3, and 4 are at 0 V. The ground plane is also at 0 V. The capacitances C_{11}, C_{12}, C_{13}, and C_{14} are the ratios V_1/Q_1, V_1/Q_2, V_1/Q_3, and V_1/Q_4, respectively. Mutual capacitance C_{32} would be the ratio V_3/Q_2.

Mutual capacitances are a function of circuit geometry. These capacitances often limit or determine circuit performance. In an integrated

---

[3] Measuring a small mutual capacitance can be difficult. One method of making a measure is to use a sinusoidal voltage at about 10 kHz and observe the current flow in a 10 k-ohm series resistor. Leakage capacitances as low as 0.1 pF can be measured this way. This capacitance measurement requires very careful shielding of the driving voltage.

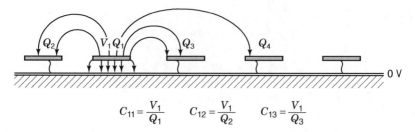

$$C_{11} = \frac{V_1}{Q_1} \qquad C_{12} = \frac{V_1}{Q_2} \qquad C_{13} = \frac{V_1}{Q_3}$$

Figure 1.8 The mutual capacitances between several traces on a ground plane

circuit amplifier, mutual capacitances are an integral part of the design. They may define circuit bandwidth and circuit stability.

1.14 DISPLACEMENT CURRENT

Figure 1.3 shows two conducting plates. If a charge Q is placed on the top plate, a charge $-Q$ must exist on the lower plate. The ratio of charge Q to the voltage on the top plate is the capacitance of this geometry. This geometry is typical of many small commercial capacitors.

If the charge stored on the capacitor plates increases linearly with time, the voltage difference V will also increase in a linear manner. A constant current source can serve to provide this increasing charge. The equivalent circuit is shown in Figure 1.9.

For convenience, we will use a standard circuit symbol for this capacitance and label it with the letter C. The current flow in this circuit can be looked at in two ways. First viewpoint: The electrons flow on to the plates of the capacitor, but they do not flow through the dielectric. Second viewpoint: In a loop analysis, the current flows through the

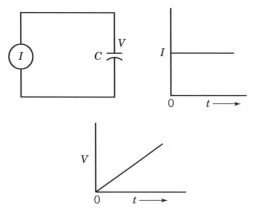

Figure 1.9 A capacitor driven from a constant current source

capacitor. Which viewpoint is correct? They both are correct providing we interpret the changing field in the dielectric correctly. As the charge Q accumulates on the plates, the D field in the dielectric also increases. This changing D field is equivalent to a current flow. We call a changing D field a displacement current. This statement is one of Maxwell's wave equations. This current has an associated magnetic field.

N.B.

A changing D field in space is equivalent to a displacement current flowing in space.

It is important to realize that we cannot have physical laws that work some of the time. Physical laws must work the same way all of the time at all frequencies. Circuit theory does not deal with the electric fields, but they are present if there are voltage differences. Later, we will discuss electromagnetic radiation. Field energy or radiation that leaves a circuit involves both an electric and a magnetic field. In a capacitor, the changing electric field is a current that has an associated magnetic field. As you will see in the next chapter, a changing magnetic field requires an electric field. In other words, these changing fields go hand in hand. In effect, energy can be carried across the capacitor plates as radiation and this requires the presence of both an electric and a magnetic field. Obviously, it is very cumbersome to view the operation of a capacitor in terms of this radiation.

1.15 ENERGY STORED IN A CAPACITOR

The energy stored in the field of a capacitor from Equation 1.8 is $\frac{1}{2} E^2 \varepsilon Ah$. If we substitute $E = V/h$ and remember from Equation 1.4 that $E/\varepsilon A = Q$, then we can write the energy E as a function of charge and voltage.

$$E = \frac{1}{2}QV. \tag{1.16}$$

We can use the ratio $C = Q/V$ in Equation 1.13 to obtain two equivalent equations for energy storage.

$$E = \frac{1}{2}CV^2 \tag{1.17}$$

$$E = \frac{1}{2}\frac{Q^2}{C}. \tag{1.18}$$

The energy that is used in fast digital circuitry must be made available near the point of demand. As you will see, it takes time to move energy over traces as multiple reflections are required. Providing local energy is the function of decoupling capacitors. This topic is discussed later in greater detail. The limitation that we will discuss is that capacitors are two-terminal devices. This means that it is impossible to take energy out and put it back in at the same time. The analogy with a water tank is appropriate, where separate input and output conduits are provided. In an oil tanker, one loading port is all that is required as loading and unloading takes place at different times. In fluid flow, a port can be a metal conduit. In electrical circuits a port usually requires two conductors. At microwave frequencies, a single metal conduit can serve as a port.

1.16 FORCES IN THE ELECTRIC FIELD

Field energy exists in the space between individual electrons. In Figure 1.1, the charges moved apart on the surface of the sphere until they were uniformly spaced. In effect they arranged themselves to store the least amount of field energy. This is characteristic of nature in so many ways. In fact all static field configurations represent a minimum of field energy storage within the geometric constraints provided. For a set of charges on conductors, there is only one field configuration possible and that field configuration stores a minimum of field energy.

N.B.

If there is a way, nature will follow a path that will reduce the amount of potential energy stored in a system. The path taken is unique. The path can involve an oscillation and usually involves some dissipation.

If the plates of a capacitor are moved closer together, the capacitance increases. Equation 1.18 shows that if the charge Q is fixed, then a larger capacitance stores less energy. This means there is a force acting on the plates trying to reduce the spacing as this reduces the potential energy stored in the system. It helps to recognize that work equals force times distance. The derivative of work with respect to distance is simply force.

Acoustic tweeters work on this principle. Parallel metal plates can be made to move air by placing an audio voltage between the plates. Potential differences in the order of several hundred volts are required. The E field must be biased at dc so that there is no doubling of frequency.

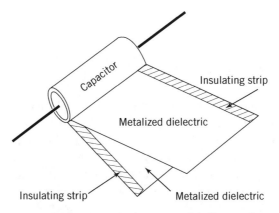

Figure 1.10   A typical wrap and foil capacitor

1.17 CAPACITORS

Capacitors are components that store electric field energy. A circuit engineer has a wide choice of component types and values to choose from. There are many construction styles, types of dielectric and voltage ratings. Capacitance values can range over 12 orders of magnitude from a few picofarads to 1 or 2 farads. Typically, electrolytic capacitors in the range 100 μF are used to store field energy for general circuit use. Capacitors in the range 0.001–0.1 μF are used on digital circuit boards for supplying local energy. These surface-mounted capacitors are often made using a ceramic dielectric called BST (barium–strontium–titanate). This material can have a relative dielectric constant as high as 10,000. In applications requiring stability over time or temperature, capacitor dielectrics can be mica or nylon. A capacitor of the wrap and foil type is shown in Figure 1.10.

Connections are made to the foil on the opposite ends of the cylinder. In many capacitor types, conductive material is vacuum deposited on a dielectric. An example of this technique is metalized Mylar®. Conductor surfaces can be made irregular to increase the effective surface area that increases the capacitance. Later, more will be said about the role capacitors play in digital circuits operating at high clock rates.

1.18 DIELECTRIC ABSORPTION

Dielectrics have a characteristic where a small charge is absorbed when an electric field is applied. In a capacitor, these trapped charges do not immediately return to the circuit when the electric field is removed. In time, these trapped electrons are released. The result can be a voltage on the terminals of the capacitor or a small current flow. This property is called dielectric absorption. This phenomenon can lead to misinformation in some measurements. Mica and nylon have a very low dielectric absorption.

The author once used a glass-encased 100-MΩ resistor as a feedback element in an amplifier. In production, the frequency response of the instrument was distorted. It was traced to a label that had been affixed to the glass. The gum backing on the label modified the parasitic capacitance associated with the resistor. Trapped charges in the dielectric were causing a small delayed current to flow in the feedback circuit.

1.19 RESISTANCE OF PLANE CONDUCTORS

There are places where large conducting surface areas must be considered. The largest is the surface of the earth or ocean, where fields cause surface currents. Sheet metal on the sides of a building provides another large surface. Compared to traces, the ground or power planes used in a circuit board design are large surface areas.

When a current flows uniformly in a conductor, the resistance is the resistivity ρ times the path length l divided by the cross-sectional area or

$$R = \frac{\rho l}{A}. \tag{1.19}$$

For a square of material where A is equal to $l \cdot t$, the resistance is

$$R = \frac{\rho}{t}. \tag{1.20}$$

Note the resistance is determined by the thickness and not the size of the square. The ohms-per-square for 1 mm thick copper at dc is 172 μΩ. For plane conductors that are not square, the resistance can be found by dividing the area into a number of squares and then using a series–parallel calculation. The resistance rises with frequency as magnetic effects limit the depth of current penetration. See the discussion of skin effect in Section 3.23.

When current does not use an entire conductor, the effective resistance rises. This is the situation at a point contact such as at a lightning strike or at a fault contact. For a copper bus bar, the contact at the ends must involve the entire surface or the end resistances will be high. In high-current applications, this can lead to overheating.

CHAPTER 2

Magnetics

OVERVIEW

This chapter discusses magnetic fields. As in the electric field, there are two measures of the same magnetic field. The H field is the direct result of current flow. The B field is the force or induction field that operates motors and transformers. As in the electric field in Chapter 1, the magnetic field is represented by field lines. The B field lines are continuous and form closed curves. The H field flux lines follow the B field lines but change intensity depending on the permeability of the material in the magnetic path.

In this chapter, the movement of electrical energy into inductors or across transformers is discussed. This extends the ideas developed in Chapter 1, where both fields are needed to move energy. Both electric and magnetic fields are needed in transformers action or to place energy into an inductor. It will be shown that iron cores in transformers reduce the magnetizing current so that transformer action is practical at power frequencies. The idea that a changing electric field creates both a displacement current and a magnetic field was discussed in Chapter 1. In this chapter, it is shown that a changing magnetic field produces both an electric field and voltages. Both fields must be in transition before electrical energy can be moved.

2.1 MAGNETIC FIELDS

We are all familiar with the magnetic field of the earth. A compass needle responds to this field to provide us with a navigational aid. We have all experimented with magnets and noted the forces that exist in the space between poles. If it were not for magnetic effects, we would not

Grounding and Shielding: Circuits and Interference, Sixth Edition. Ralph Morrison.
© 2016 John Wiley & Sons, Inc. Published 2016 by John Wiley & Sons, Inc.

have motors, generators, or transformers. These devices are basic to our technology. It is easy to forget that magnetic fields are at the heart of operating all of our electronic circuitry. The story starts again with the atom.

In a few elements, the atomic structure is such that atoms align to generate a net magnetic field. Magnetite is a mineral containing iron that can produce a magnetic field that can deflect a compass needle. Rare earths neodymium and samarium, iron, cobalt, and nickel plus ceramic fillers are used to make magnetic materials that support or generate magnetic fields. Silicon steel is used in many motors, generators, and power transformers. Permanent magnets using rare earth elements develop strong B fields that make speakers practical in our cell phones and provide efficient drives for electric autos.

The flow of electrons is another way to generate a magnetic field.[1] These electrons can be flowing on a conducting surface, along a circuit trace or in free space. The simplest geometry to consider is a long cylindrical conductor. To demonstrate that a current generates a magnetic field, thread a conductor through a piece of paper. Allow a dc current to flow in the conductor. Iron filings placed on the paper will line up in concentric circles around the conductor. A small compass needle in the vicinity of the conductor will also align itself with the magnetic field pattern.

The letter H is reserved for the magnetic field generated by a current. Figure 2.1 shows the shape of the H field around a long, straight conductor carrying a direct current I. As in the electric field, lines are used to indicate the shape of the field.

In Figure 2.1, the field lines form circles around the conductor. In this representation, the closer the field lines, the more intense the field. These field lines are also called lines of magnetic flux. Note the field intensity for this geometry is constant along a path that is concentric with the conductor.

The field intensity (flux density) is equal to the number of lines that cross a unit area perpendicular to the lines. Note the magnetic field is a vector field. At every point in space, the field has an intensity and a direction.

The magnetic field is a force field. This force can only be exerted on another magnetic field. If two parallel conductors carry current in the same direction, the resulting force will try to move the conductors

---

[1] To be completely correct, it should be stated that a changing D field is equivalent to a displacement current. This current also creates a magnetic field. This is one of Maxwell's equations.

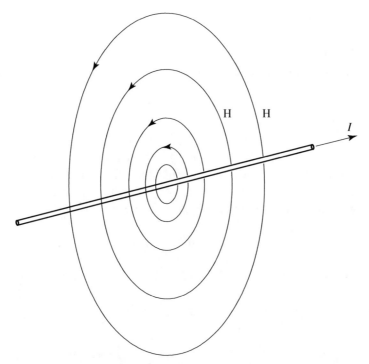

Figure 2.1   The H field around a current-carrying conductor

together.[2] The direction of the force, the direction of the current flow, and the direction of the field lines are all perpendicular to each other.

2.2 AMPERE'S LAW

Ampere's law states that the integral of the H field intensity in a closed-loop path is equal to the current threading that loop.

$$\oint H \; dl = I. \tag{2.1}$$

The simplest path to use for this integration is one of the concentric circles in Figure 2.1, where H is constant and r is the distance from the conductor. Solving for H, we obtain

$$H = \frac{I}{2\pi r}. \tag{2.2}$$

---

[2] The force is in the direction to reduce the energy stored in the circuit inductance. See Section 2.6.

From this equation, we see that H has units of amperes per meter. In this geometry, the H field falls off linearly with distance. The value of H is constant at a distance *r* from the conductor. For a long conductor, the H field falls off linearly with distance.

2.3 THE SOLENOID

The magnetic field of a solenoid is shown in Figure 2.2. Note that the field intensity inside the solenoid is nearly constant while outside the solenoid, the field intensity falls off very rapidly. Using Ampere's law, the integral of H in the solenoid is approximately

$$\oint \mathbf{H} dl \cong nI \ l. \tag{2.3}$$

The full integral is not easily performed as H is not constant around the entire path. The important thing to note is that the H field in the solenoid is proportional to the current and to the number of turns in the solenoid.

2.4 FARADAY'S LAW AND THE INDUCTION FIELD

When a conducting coil is moved through a magnetic field, a voltage appears at the open ends of the coil. This is illustrated in Figure 2.3. The voltage depends on the number of turns in the coil and the rate at which

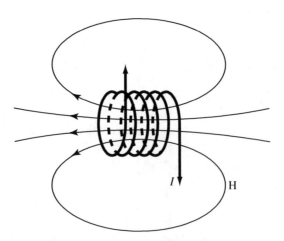

Figure 2.2   The H field around a solenoid

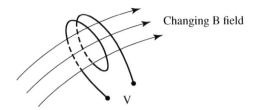

Figure 2.3   A voltage induced into a moving coil

the flux is changing that threads the coil. No voltage results if the coil is moved so that the total flux through the coil remains constant.

The magnetic field has two measures. This is similar to the electric field that has an E and a D field measure. We have already considered the H or magnetic field that is proportional to current flow. The force field representation that induces voltage is called the B or induction field. The relation between B and H fields is given by

$$B = \mu_R \mu_0 H \qquad (2.4)$$

where the factor μ_0 is the permeability of free space and μ_R is the relative permeability of the medium. Air has a relative permeability near unity. The factor μ_0 is equal to $4\pi\ 10^{-7}$ T/A m (teslas per ampere-meter). The relative permeability of iron can vary from 500 to 100,000. For an area of constant field intensity, the magnetic flux ϕ is simply the product BA where B is in tesla, A is the area in square meters, and ϕ is the flux in webers. We shall refer to this flux as "lines." The voltage induced in a conducting coil is

$$V = n\left(\frac{d\phi}{dt}\right) \qquad (2.5)$$

where n is the number of turns in the coil. This equation can be written in terms of the B field as

$$V = nA\ \frac{dB}{dt}. \qquad (2.6)$$

In a magnetic structure where a steady voltage is applied to a coil, the B field intensity increases linearly. The lines representing the B field form continuous loops that pass through the coil. The lines representing the H field are parallel to the B field lines, but the H field intensity varies inversely to the permeability. Equation 2.4 gives the relation between B and H.

Equation 2.5 is known as Faraday's law. If the induction flux B increases linearly, a steady voltage V must exist at the coil ends. The inverse is also true. If there is a fixed voltage on this coil, then the B field intensity in the coil must be increasing linearly. This is known as Lenz's law. These laws are very difficult to observe for simple coils in air at frequencies below a few megahertz.

In electrostatics, the E field is the force field. In magnetics, the B field is the force field. The force on a current loop in a magnetic field is proportional to the current and the B measure of the field produced by the current.

2.5 THE DEFINITION OF INDUCTANCE

Inductance is defined as the ratio of magnetic flux generated per unit current. It is a difficult problem to calculate this total flux in a typical geometry. A practical way to measure inductance is to use Faraday's law. We begin by looking at the coil in Figure 2.2. Equation 2.6 states that the magnetic field B will increase at a constant rate if a steady voltage is applied to the coil. Both the H field and the induced voltage are proportional to the number of turns. Therefore, the voltage V is proportional to n^2. For a coil in air, $B = \mu_0 H$ and Equation 2.6 can be rewritten in terms of a changing current as

$$V = n^2 A k \mu_0 \left(\frac{dI}{dt} \right) = L \left(\frac{dI}{dt} \right). \tag{2.7}$$

where k relates to the geometry of the coil. The factor $n^2 A k \mu_0$ is the inductance L of the coil. The unit of inductance is the henry. Equation 2.7 states that if $V = 1$ V, then for an inductance of one henry the current will rise at the rate of 1 A/s. A henry is a large unit of inductance. Typical circuit inductors range from a few microhenries to a few millihenries. These units are abbreviated as mH and μH. In digital structures, when we discuss decoupling capacitors, inductances in the picohenry range will be important.

As mentioned earlier, the inductance in Equation 2.7 is for a coil in air. Most commercial inductors are constructed by associating the coil with a magnetic material. We will discuss this construction in the following sections.

2.6 THE ENERGY STORED IN AN INDUCTANCE

In the electric field case, work was related to moving a small test charge in the electric field. In the magnetic field case, a test magnetic field (unipole) does not exist. Pushing on a small charged mass in a magnetic field causes the mass to move at a right angle to the direction of the force, and this is not a measure of work. One way to calculate the work stored in a magnetic field is to use Equation 2.7. The voltage V applied to a coil results in a linearly increasing current. At any time t, the power P supplied is equal to VI. Power is the rate of change of energy or $P = dE/dt = \text{V}I$, where E is the stored energy in the inductance. Since the voltage $\text{V} = LdI/dt$, the stored energy in an inductance L is

$$E = L \int_0^I I dI = \frac{1}{2} L I^2. \tag{2.8}$$

N.B.

An inductor stores field energy. It does not dissipate energy.

The presence of a voltage V on the terminals of an inductor implies an electric field. The movement of energy into the inductor thus requires both an electric and a magnetic field. This is very similar to what we found when we placed field energy into a capacitor. While moving charge into the capacitor, we created a magnetic field. We have just shown that both the E and B fields must be present to move energy into an inductor. To remove energy from an inductor or a capacitor, both fields must again be present.

Faraday's law requires a voltage when a changing magnetic flux couples to a coil. This voltage means electric field energy must be present. This energy is stored in a distributed manner between every conducting element pair on the coil. When a steady current flows in the inductor, the magnetic flux is constant. This means that the voltage is zero and there is no electric field energy stored. When the circuit is opened, the current starts to decrease. The result is a changing magnetic flux that creates a voltage, and this voltage begins to place energy into the interwinding capacitance of the coil. In effect the current continues to flow, but now it flows in the distributed capacitance of the coil. The magnetic field energy begins its conversion to electric field energy.

N.B.

The energy stored in an inductor cannot simply vanish. It must go somewhere.

The field energy in the inductor is $1/2 LI^2$. The field energy in a capacitance is $1/2 CV^2$.

Consider a 1-mH inductor carrying a current of 0.1 A. Assume the shunt capacitance equals 100 pF. The stored energy is 5×10^{-4} J. When this energy is fully transferred to the capacitance, the voltage must be 3116 V. The natural frequency of the inductance and its own parasitic capacitance is about 500 kHz. The energy transfers from the inductor to the capacitor in $1/4$ cycle or in 0.5 μs. Mechanical contacts cannot open very far in 0.5 μs, so the result is the voltage breakdown of air. For a relay contact, this sudden rise in voltage results in an arc. The energy that was stored in the inductor now goes into light and heat. If the switch is a semiconductor, the resulting voltage would probably destroy the component. There are several ways to absorb the stored magnetic field energy and avoid a high voltage. A reverse diode across the coil can provide a path for interrupted current flow. Another technique is to place a capacitor across the inductance. This will lower the natural frequency and at the same time reduce the voltage.

N.B.

The field energy in an inductance cannot be dissipated in zero time.

2.7 MAGNETIC FIELD ENERGY IN SPACE

To solve for the magnetic field energy stored in a region of space, we can invent an increment of magnetic field from an increment of current flowing in a closed superconducting loop. We can use this increment of magnetic field to add field energy into a main superconducting loop. When the incremental loop of current is moved a distance d in the B flux of the main loop, the work done on the main loop is $BHAd$, where HA is the flux from the incremental current loop. This work increments the current in the main loop, which increases the field intensity. To build up the field energy in the main loop, we must bring the energy across in small increments. At the beginning, no work is required as the initial B field is zero. When the B field is maximum, the work per unit current is $W = BHA$. The average work done to store energy in the B field is $1/2$ this value. Since $B = \mu_0 H$ in a vacuum, we can write the energy stored as

$$E = \frac{1}{2}\left(\frac{B^2 V}{\mu_0}\right) \tag{2.9}$$

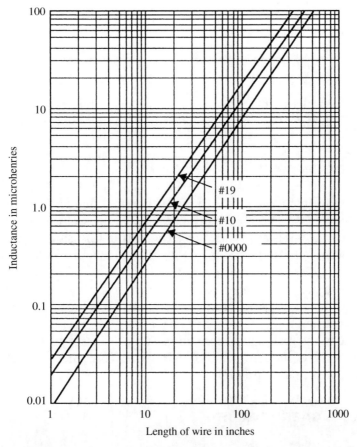

Figure 2.4   The inductance of round copper conductors

where the volume $V = Ad$ and μ_0 is the permeability of free space.

Space can store magnetic field energy. In this sense, every volume of space has an inductance. This is the counterpart to the electrostatic case, where every volume in space has a capacitance.

There is magnetic flux around an isolated conductor carrying current. The flux per unit length defines an inductance per unit length. Figure 2.4 shows the inductance of round copper wires as a function of length. The important thing to appreciate is that this inductance is essentially independent of conductor diameter. At a frequency of 10 MHz, a 20-in.-long #19 conductor looks like 120 Ω. A 20-in.-long #0000 conductor looks like 48 Ω. Three parallel #19 wires spaced a few inches apart would look like 30 Ω. It is obvious that the amount of copper is not the issue. The geometry of the copper makes all the difference. At high frequencies, it is essentially impossible to short two points together. This important point will be discussed later in the book.

N.B.

Heavy conductors are not the solution to limiting potential differences. The reason is simple. They do not get rid of fields.

2.8 ELECTRON DRIFT

Current flow in conductors is the movement of charge. The velocity of energy flow is the speed of light, but the average velocity of electrons in a typical circuit is extremely low. The reason this is true is that every copper atom in a conductor can contribute an electron and the number of atoms follows from Avogadro's number, which is 6.022×10^{23} atoms per gram-mole. In a typical circuit, conductor carrying current, the average electron velocity is less than 0.001 in./s. This is further reason to use fields to explain the movement of energy.

A note to the reader: The remaining subjects in this chapter deal with the magnetics of inductors and transformers. You can skip to Chapter 3 without a loss in continuity.

2.9 THE MAGNETIC CIRCUIT

Magnetic materials are needed to build motors, generators, and transformers. These materials are also used to make practical inductors. To understand the role of magnetic materials, we start with Figure 2.5, where a conducting coil is wound on a simple toroid made of magnetic material. Assume a steady voltage V is applied to the coil starting at time $t = 0$.

From Equation 2.5, the voltage controls the rate of change of the B field or $d\text{B}/dt = \text{V}/nA$. The maximum B field intensity[3] in a typical power transformer core is about 1.5 T (15,000 gauss). If V is a constant, we can calculate how long it will take the field intensity to increase to this level. Since $\text{B} = t\text{V}/nA$, where A is the area and V is the voltage, the time t for B to reach 1.5 T is

$$t = \frac{nA\text{B}}{\text{V}} = 1.5 \ nA/\text{V}. \tag{2.10}$$

---

[3] The unit of magnetic induction used by transformer designers is the gauss. One gauss is 10^{-4} T.

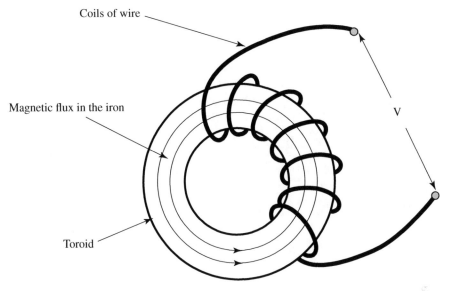

Figure 2.5 A coil wound on a toroidal core of magnetic material

If the area A is 1 cm^2 (10^{-4} m^2), V = 1 V, and n = 100 turns, the time t is 0.015 s. This time is independent of the type of magnetic material in the core.

The H field associated with this B field can be determined by Ampere's law. If we assume the H field is constant in the toroid and if we choose a closed path that threads all of the turns of the coil, then

$$H\,dl = 2\pi r H = nI \qquad (2.11)$$

$$I = \frac{2\pi r H}{n}. \qquad (2.12)$$

To calculate I, we need to relate B to H in the magnetic material. Using Equation 2.4,

$$I = \frac{2\pi r B}{n\mu_R\mu_0}. \qquad (2.13)$$

If we assume $r = 0.1$ m, $\mu_R = 1$, and setting $\mu_0 = 4\pi 10^{-7}$, the value of I is 7500 A.

Obviously, this level of current is not acceptable. If the relative permeability of the core material is 50,000, the current reduces to 0.15 A, a very practical number.

The current required to supply the B field is called a magnetizing current. Without the presence of a magnetic material, this example shows that at power frequencies an air transformer is of no practical use. If the permeability of the material were infinite, the magnetizing current would be zero. An ideal transformer requires no magnetizing current.

It is interesting to calculate the energy stored in the magnetic material and to determine the inductance of this geometry. Assume the volume V of the toroid material is $2\pi 10^{-5} \mathrm{m}^3$. Assume there is no gap. Now set B maximum equal to 1.5 T. Using Equation 2.9 and setting $\mu_0 = 4\pi\,10^{-7}$, if $\mu_R = 50{,}000$, the energy E is $4.5 \times 10^{-3}\,\mathrm{J}$. The inductance using the relationship $E = \frac{1}{2}\,LI^2$ is 0.4 H. This is a large inductance value that can store a very limited amount of energy.

2.10 A MAGNETIC CIRCUIT WITH A GAP

We next consider the effect of placing an air gap in the magnetic path in Figure 2.5. This is shown in Figure 2.6.

When a steady voltage V is applied to the coil, the B field will increase by Faraday's law. The gap has no influence on the buildup of the B field. Remember the B field is continuous around the magnetic path. The H

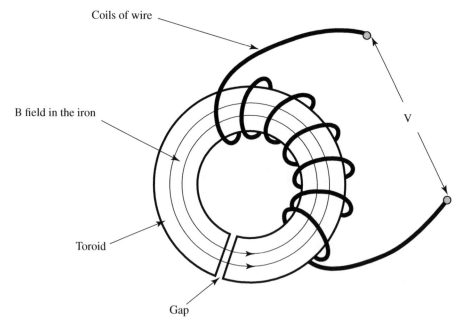

Figure 2.6   A magnetic circuit with an air gap

field in the magnetic material is $B/\mu_0\mu_R$. In the gap, the H field is B/μ_0. Note that the H field is much larger in the gap.

The H field in the gap must be B/μ_0 or 1.5 T divided by $4\pi \times 10^{-7}$ or 1.19×10^6 A/m. For 100 turns, the current requirement is reduced by a factor of 100. The current required to establish this H field in a gap of 10^{-3} m is 11.9 A. The current required to establish the H field in the magnetic material is essentially the same as before or 150 mA. The total current is therefore 12.05 A.

The energy stored in the gap is $\frac{1}{2}B^2V/\mu_0$, where the gap volume is 1 cm$^2 \times 0.1$ cm $= 10^{-5}$ m^3. This energy is $1.12 \times 10^{-5}/4\pi \times 10^{-7} =$ 8.75 J. Without a gap the energy stored was 4.5×10^{-3} J, a small number. The inductance using Equation 2.8 is 0.241 H, a significant number.

N.B.

Magnetic field energy is largely stored in air not in a magnetic material. Without an air gap, the energy that can be stored is very small.

N.B.

For a geometry involving air and dielectrics, the majority of electric field energy is stored in the air. See Figure 1.7.

The shape of the magnetic material serves to focus the B field into the gap. The magnetic field follows the permeable material as this path takes so much less energy than using the nearby air space. This is another example of nature configuring a field to store the least amount of energy. In this example, the B field is near maximum. If the magnetic material becomes saturated, the ability to store energy is lost. When saturated, the magnetic material has a relative permeability of 1.

The toroid with a gap in Figure 2.6 is called an inductor. The quality of the inductor at different frequencies depends on the type of magnetic material and how the coil is wound.

2.11 SMALL INDUCTORS

Inductors in the microhenry range are often wound on a small cylindrical core. The flux path is in the space around the inductor. Since very little energy is stored in a magnetic material, the energy storage for these inductors is in the surrounding space.

A conductor that is threaded one or more times through a ferrite bead forms an inductor. The inductance is in the order of nanohenries. One nanohenry at one gigahertz is only 6.28 Ω. This level of inductance is only effective in very low-impedance circuits. If there is any low-frequency current in the conductor, the core may be saturated. This can occur for utility power. The presence of the core may space the conductors thus reducing cross coupling. In this case, the magnetic material is not needed. It is important to note that the permeability of magnetic materials falls off rapidly above a few megahertz.

2.12 SELF- AND MUTUAL INDUCTANCE

The magnetic flux generated by a current can couple into nearby circuits. The flux associated with this coupling is called leakage flux. The ratio of the leakage flux that couples into a second circuit to the generating current is called a leakage or mutual inductance. The symbol L_{12} represents the flux coupled to circuit 2 from a current flowing in circuit 1. The symbol L_{11} represents a self-inductance. The inductor in Figure 2.6 has a self-inductance of 0.241 H. In this case, all the flux of the inductor couples to its own coil.

In Figure 2.6, there is an H field close to the current-carrying conductor. This is H field leakage flux that does not use the core as the magnetic path. This means that circuits that are very close to the toroid can couple to this field. Around the gap, the leakage flux can be more substantial. Many inductors are wound using "cup cores." In this geometry, the gap is in the very center of the core and the leakage flux is limited to a small value.

The circuit symbol for inductance implies that the field energy is stored inside the inductor. Small-valued inductors (microhenries) often store a fraction of their field energy in the space around the component. Some of this field radiates and is not returned to the circuit. This is a situation not covered by circuit theory. We will discuss this radiation in a later chapter.

2.13 TRANSFORMER ACTION

When a steady voltage is connected to the coil in Figure 2.5, the B flux increases linearly with time. From Equation 2.10, if $n = 100$ turns and the voltage is 1 V, the B field will reach an intensity of 1.5 T in 15 ms. If the voltage is 10 V, the time would be 1.5 ms. When the B field reaches 1.5 T, we can reverse the voltage polarity. After this reversal, the flux intensity starts decreasing. In another 1.5 ms, the field intensity is again zero. After another 1.5 ms, the flux intensity is −1.5 T. At this point if there is a second voltage polarity reversal, the flux would again return to zero in an additional 1.5 ms. This full cycle would take 6.0 ms. If this cycle is repeated over and over, the resulting waveform is called a square wave at a frequency of 166.6 Hz. The B field we have discussed is independent of the permeability of the magnetic material in the core.

The flux pattern and the coil voltage are shown in Figure 2.7.

The square wave voltage in Figure 2.7 has no dc component. In theory if a dc component were applied, the core would saturate in a few cycles. In a practical situation, a slight dc offset can be accommodated. The penalty is an asymmetrical flow of magnetizing current and some distortion in the voltage waveform.

N.B.

A magnetic material is needed to limit magnetizing current.

We now consider a core with two coils as shown in Figure 2.8. The two coils are called the primary and the secondary windings, respectively.

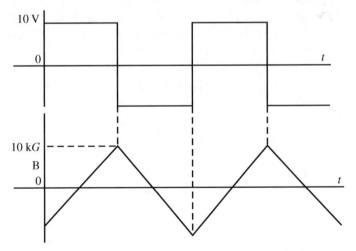

Figure 2.7 The flux pattern for a square wave voltage applied to a transformer coil

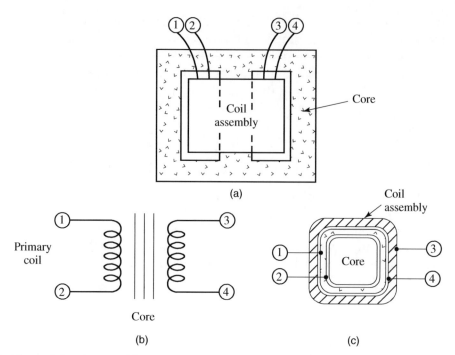

Figure 2.8   A core with two coils that form a simple transformer. (a) Assembly, (b) symbol, and (c) winding arrangement

If a voltage is applied to the primary coil, the B field, we have just described, couples to both coils. An oscilloscope placed on the second coil would show a square wave at 166.6 Hz. Since the original square wave was 20 V peak-to-peak, the secondary voltage would be the same. If the number of turns on the secondary were increased to 200, the voltage would be 40 V peak-to-peak. If the number of turns on the primary and secondary were both doubled, the voltages would not change. The differences would be that the maximum B field would be 0.75 T. Since the magnetizing current is proportional to the H field and inversely proportional to the number of turns, the magnetizing current would be reduced by a factor of 4.

If a load resistor is placed on the secondary coil, the current that flows by Ohm's law is V/R. To keep the net H field constant, this same number of ampere-turns must flow in the primary coil. If there are 100 primary turns and 200 secondary turns, a secondary current of 0.1 A requires a primary current of 0.2 A. The current in the primary coil must now equal the magnetizing current plus the load current. In a well-designed transformer, the magnetizing current is held to a few percent of the maximum load current. It is convenient to accept the idea that the magnetizing current creates the B field in the magnetic material.

The energy stored in the magnetic material is stored in what is called the magnetizing inductance. Ideally, current flowing in an inductance produces no heat. There is some heat loss because the magnetizing current flows in the primary coil resistance. There are eddy current losses in the magnetic material as there are currents associated with a changing magnetic field. See Section 2.14. Consider the heat loss in a 10 kW transformer. A 1% loss (100 W) would be a good design. This amount of heat can cause a significant rise in temperature in the center of the transformer. Usually, some form of forced ventilation is required for this size of transformer.

In a transformer, secondary load impedances are reflected to the primary side multiplied by the turns ratio squared. In the example above, the turns ratio is 1:2. A 100-Ω load on the secondary appears as a 25-Ω load to the primary. To show this is true consider a 10-V secondary voltage and a current level of 0.1 A or a power level of 1 W. On the primary, the voltage is 5 V and the current is 0.2 A. The primary voltage source sees a resistance of 5 V/0.2 A = 25 Ω.

A practical transformer has series leakage inductance and shunt capacitance associated with coil construction. These reactances appear as loads to any primary voltage.

Reactive loads are reflected across the transformer by the turns ratio squared. This means that the secondary leakage inductance is multiplied by the turns ratio squared and the secondary capacitance is divided by this turns ratio squared. The equivalent circuit of a transformer is shown in Figure 2.9. The transformer symbol in the figure represents an ideal transformer with a step-up turns ratio of 1:n.

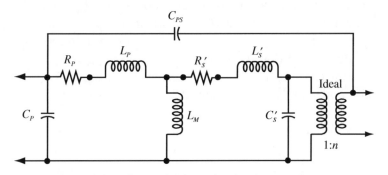

Figure 2.9 The equivalent circuit of a transformer

The magnetizing inductance is labeled L_m. The resistances of the coils are labeled R_P and R_S, respectively.

There are capacitances associated with the primary and secondary coils labeled C_P and C_S. The inductances associated with the leakage flux are labeled L_P and L_S. The primes indicate that the values have been corrected for the turns ratio squared and referenced to the primary side of the transformer. The secondary leakage inductance and resistance are divided by n^2. The secondary shunt capacitance is multiplied by n^2.

The load current and the magnetizing current both flow in the primary leakage inductance. Only the load current flows in the secondary leakage inductance. This means that the leakage flux is load dependent. In facilities with large distribution transformers, load currents can be hundreds of amperes. The leakage flux from these transformers can interfere with some nearby electronic devices. It is possible to limit the leakage inductance in the design of a transformer by interleaving the primary and secondary coils and by using more core material that allows fewer turns per volt.

The shunt capacitances associated with transformer coils are related to the storage of electric field energy. If there are conductors and a voltage difference, there must be an E field. Every conducting element along the conductor stores field energy with every other conducting element depending on separation and voltage difference. Winding a coil so that the starting turns are near the ending turns will greatly increase the stored electric field energy. In some power transformers, the start and end points are separated to avoid a possible voltage breakdown. Most of the field energy is stored in air, so the parasitic capacitances are not greatly influenced by a dielectric. The capacitances between the primary and secondary coils pose a complex problem that will be discussed later. In Figure 2.7, this mutual capacitance is represented by C_{12}.

The magnetizing inductance of a transformer can be measured at the primary connections if the secondary coils are open circuited. The result will vary depending on voltage level and the frequency selected. The leakage inductance can be measured from the primary winding if the secondary coils are all short circuited. The test voltage must be a small fraction of the normal operating voltage.

When a voltage is impressed on the primary coil of a transformer, there is an electric field. As the current builds up, there is a magnetic field. The transfer of energy across the transformer requires the presence of both fields. As we have seen, placing energy into a capacitor or

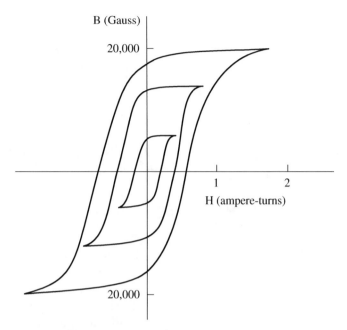

Figure 2.10   Typical hysteresis curves

into an inductor requires both fields. A transformer requires both electric and magnetic fields to transfer energy from the primary side to the secondary side.

2.14 HYSTERESIS AND PERMEABILITY

The relationship between the B and H fields in a magnetic material is not linear. The B/H curves for a typical material are shown in Figure 2.10. The ratio of maximum B to maximum H is one measure of permeability. Typically, B/H curves are shown for a sinusoidally varying B field. Note that this measure of permeability varies with B maximum. B/H characteristics will also vary with frequency and with primary voltage waveform. This means that a permeability figure is an approximate measure for a complex relationship.

There are a large number of magnetic materials for use in magnetics. High silicon transformer steel does not have a high permeability for small values of B. Mumetal®, on the other hand, has a very high permeability at low flux densities. Mumetal is difficult to use and is expensive. Ferrite materials have excellent permeability at high frequencies. Manufacturers of magnetic materials provide the designer

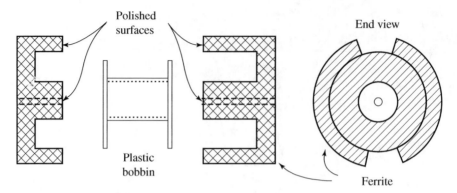

Figure 2.11   Ferrite cup-core construction

with hysteresis curves that are related to the expected uses of that material.

For a sinusoidal voltage, the nonlinear relation between B and H means that the magnetizing current is not sinusoidal. As a magnetic material approaches saturation, the demand for magnetizing current increases significantly. For a small power transformer, the *IR* drop from magnetizing current can distort the voltage waveform on the primary coil. In this case, neither the B nor H field is sinusoidal.

In a power transformer operating at 60 Hz, the point of maximum flux density occurs when the voltage is at a zero crossing. If there is core saturation, it will occur at these voltage crossings. This timing can be used as a signature for this class of interference. If there is voltage distortion at the peak of voltage, the problem is usually related to the peak current demanded by rectifier circuits.

2.15 EDDY CURRENTS

Consider a closed circular path inside a conducting magnetic material. If the field crossing this loop changes, a voltage must result. This voltage in the material results in eddy current flow that dissipates heat. This induced current flows in circles around the changing lines of flux. One technique to limit this current flow is to construct the magnetic core from thin insulated laminations. Typically at 60 Hz, 15-mil laminations are used and at 400 Hz, the laminations are 6 mils thick. These thin laminations break up these eddy current paths without significantly reducing the effective core area.

For transformer applications above 400 Hz, the preferred core material is ferrite. A ferrite is made from finely powdered magnetic material suspended in a ceramic filler. When the mixture is fired, the resulting material is similar to a ceramic insulator. The eddy current losses in this type of magnetic material are quite low. Dc-to-dc converters use ferrite core transformers because of their excellent high-frequency characteristics. Figure 2.11 shows a typical ferrite cup-core arrangement.

Transformer coils are wound on a bobbin that is mounted on the center leg of the core. Cup cores can also be supplied with a built-in center gap. This configuration is ideal for building inductors. As we have seen, a gap reduces the effective permeability of the core but provides for energy storage. In transformer applications, a gap is not desired. To remove the gap, the mating surfaces of the core cups are carefully machined and polished. These cups are always supplied as mating pairs.

CHAPTER 3

Digital Electronics

Buildings have walls and halls.
People travel in the halls, not the walls.
Circuits have traces and spaces.
Energy travels in the spaces, not the traces.

— Ralph Morrison

OVERVIEW

This chapter shows that both electric and magnetic fields are needed to move energy over pairs of conductors. The idea of transporting electrical energy in fields is extended to traces and conducting planes on printed circuit boards. Logic signals are waves that carry field energy between points on the board. These waves are reflected and transmitted when different transmission lines are interfaced. There are several sources of first energy that play a role in circuit performance. These sources are connected logic, the ground/power plane structure, and decoupling capacitors. Decoupling capacitors are actually short stub transmission lines that supply energy.

The use of vias in the transmission paths is discussed in detail. The fact that energy cannot pass through a conducting plane is stressed. Limiting interference coupling in an A/D converter is a problem in keeping analog and logic fields separated. Terminating balanced transmission lines is also discussed.

The concept of displacement current and its associated magnetic field is important. These ideas show how field energy flows into a transmission line and is placed into capacitance at the leading edge of the wave. Radiation occurs at the leading edge of a wave as it moves down the transmission line.

Grounding and Shielding: Circuits and Interference, Sixth Edition. Ralph Morrison.
© 2016 John Wiley & Sons, Inc. Published 2016 by John Wiley & Sons, Inc.

3.1 INTRODUCTION

This chapter covers material that might be considered the analog aspect of digital design. This includes conductor geometry but not logic or software. Circuit boards use multiple ground and power planes intermixed with traces to interconnect memory, logic, microprocessors, optics, power, and data streams as well as analog components. Successful layouts require an understanding of how transmission lines move energy stored in decoupling capacitors to components mounted on multilayer circuit boards. These waves carry energy to operate components and they also transport both logic and interference. Keeping these functions separate is the job of the circuit board designer.

3.2 THE TRANSPORT OF ELECTRICAL ENERGY

Circuit theory suggests that conductors carry energy. Conductor geometry is rarely shown in schematics. So far, we have discussed voltages and currents, and their associated E and H fields. We have seen that fields both store and move energy, and that conductors serve to direct the path of energy flow. We have seen that both fields are necessary to move energy into or out of a capacitor or inductor, or to transfer power across a transformer. The two fields are present for a flashlight at dc or for a power distribution system at 200 kW. The fields associated with a flashlight are shown in Figure 3.1.

Nature always seeks a way to distribute energy so that there is less potential available to do work. As an example, water will always run downhill if there is an opportunity. Field energy seeks a configuration that stores the least energy for the conductor geometry involved. Nature takes this course of action in every system. We take advantage of this characteristic to make electromagnetic fields do our bidding in a circuit. We take energy from the power grid and store it in capacitors. Then we let circuits send packets of this stored energy to components to cause

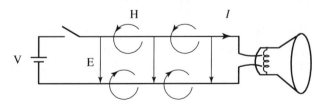

Figure 3.1 The electric and magnetic fields associated with a flashlight

some logical end result. We play a game using nature's rules to accomplish our tasks.

In this chapter, the subject of ac power, power transformers, and regulating dc power supplies will not be discussed. We will use an ideal voltage source that can supply any amount of energy without sagging. We will use conductors with zero resistance. Power supplies are a part of analog design and this is the subject of Chapter 4.

Information can take the form of a change in voltage. A change in voltage is a change in an electric field. A changing electric field implies a displacement current and an associated magnetic field. The magnetic field can be very small, but it must be present if information is to be moved. This movement of information is the movement of an electromagnetic field. A field cannot be created or moved in zero time as this would require infinite power. This movement of fields using conductors is the subject of this chapter. All electrical activity involves moving fields. In digital circuits, transmission line theory describes the movement of energy and we interpret the resulting voltages as logic. When waves are in motion, they move at the speed of light. In air, this speed is 300 m/μs.

N.B.

The movement of electrical information requires moving electrical energy, which requires changing electric and magnetic fields.

The material in this chapter is directed at the flow of digital information on circuit boards. Most of this information is applicable to the flow of electrical energy in analog and power circuits. For example, power conductors can act as transmission lines and carry power as well as rf energy into analog circuits.

3.3 TRANSMISSION LINES – INTRODUCTION

Transmission line theory was an important tool used by radio engineers to handle the flow of energy from transmitting tubes to a transmitting antenna. In the early days of electronics, these transmission lines were often just pairs of open wires. The goal in design was to avoid reflections in the transmission line and maximize radiation from the antenna. The signals that were transported were sinusoidal. In digital circuits, our interests are in logic signals that are called step voltages. We will also call these moving signals waves. We will move these waves over transmission lines usually built on circuit boards.

Transmission line theory is often presented using lumped parameter symbols that represent a distributed inductance and capacitance for the conductors. Traces on a circuit board are short transmission lines. Trace pairs, traces over a conducting plane, or traces between two conducting planes are all transmission lines. The object in circuit board design is to concentrate fields under or between traces so they have a uniform character and to insure that very little field leaves the board. An examination of field patterns shows that a small amount of field can leave each transmission line at the edges and over the tops of traces. Because there can be hundreds of traces on the outer surfaces of a board, this can be a source of circuit board radiation.

Why does field energy follow pairs of conductors? The answer is very simple. It is easier for energy to follow these paths and then to jump into space. For power transmission at 60 Hz, very little energy leaves the space between the conductors and radiates. The fields follow the conductors wherever they go. At 400 Hz, the distribution of power on open wiring is limited to a few hundred feet. Above this frequency, open power wiring is not practical. Above 400 Hz, power must be transported using coax to limit losses. Above about 500 kHz, antennas can easily radiate energy. Radiating antennas have been built that work at frequencies as low as 10 kHz. The antenna dimensions here might exceed 8 miles. Our interests on a circuit board are usually to avoid radiation. Our aim is to control the path taken by fields. The spectrum we will consider goes from dc to many gigahertz.

The energy flowing from a source to a load flows in fields. When a new load is added, a series of waves must propagate back and forth between the load and the source until the new power level is set. The changing field (wave action) in the space between the source and the load can be a source of interference. By Faraday's law, circuits that share the same space as the source circuit will couple to this changing field. This interference can be avoided if the field from the source is confined to a small volume in space. Without a field view of energy transport, this coupling process would not be apparent. This is exactly the situation in an electrical system. Demands for power are transported in fields. If these requests have short rise times, the changing field can introduce interference into nearby cables that can carry the interference into electronic hardware. Before we discuss interference, we need to examine in more detail how energy is transported on transmission lines.

N.B.

Fields transport electrical energy at all frequencies including dc.

3.4 TRANSMISSION LINE OPERATIONS

Consider the battery, switch, and transmission line in Figure 3.2. This lumped parameter circuit might represent parallel traces on a circuit board. This is an ideal line with no losses. For this discussion, consider that the line extends to infinity. At the moment the switch closes, a voltage and its associated E field appears across the start of the traces.

In the first increment of time, a charge flows into the first increment of line capacitance. This flow of charge is a current that creates a unit of magnetic field associated with the first increment of inductance. This increment of inductance involves both the outgoing and return conductor. The magnetic field uses the entire space between two conductors even though this representation seems to imply two fields. In the second increment of time, the second increment of capacitance receives charge and the next increment of magnetic field is generated. After the switch closes there is a steady current as the same amount of charge is supplied for each increment of time. The net effect is a wave of field energy traveling down the transmission line filling the space with E and H fields. In a typical trace over a ground plane on a circuit board, the wave velocity is about one half the speed of light. This velocity is related to the fact that on a typical circuit board, the field energy flows in a dielectric between two conductors. Fields flow slower in a dielectric than in space.

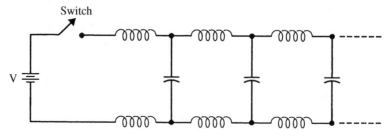

Figure 3.2　A battery, a switch, and a transmission line

Any number of waves can use a transmission line at the same time. It is useful to idealize a wave as having a steep leading edge followed by an unchanging voltage. At the leading edge, the slope of the changing voltage controls the magnitude of the displacement current that flows into the capacitance of the line. I avoid using a step voltage to represent a wave as this would imply an infinite displacement current that cannot exist. Behind the leading edge, a steady current flows on the circuit traces. This current flows in the inductance of the transmission line creating an H field. Thus, behind the leading edge, there is a steady E and H field. This moving energy fills the space behind the wave front. The only motion we can detect behind the wave front is this current flow.

In a transmission line, a passing wave places energy into the distributed capacitance but only when and where the E field is changing. Remember that D = εE. A changing D field is a displacement current. This displacement current at the leading edge of the wave completes the current path between the two conductors. The battery current goes out on the top conductor, through the leading edge of the wave and returns on the bottom conductor. If the leading edge is stretched out in space, the displacement current is also stretched out in this same space. In the volume of space associated with the leading edge, the moving energy is converted to E field static energy that is stored in the capacitance of the line.

A fixed voltage source and a steady current flow imply that the transmission line looks like a resistance. This resistance is called the characteristic impedance of the line. The term impedance in circuit theory implies sine waves and phase shift. This meaning is generally ignored in logic. Of course, the term impedance might take on its proper meaning when sine wave signals are transported between trace pairs. Even though the transmission line has a characteristic impedance that has units of ohms, there are no losses in this resistance.

In logic, the term characteristic impedance applies to the geometry of the transmission lines. This impedance Z is equal to

$$Z = \left(\frac{L}{C}\right)^{1/2} \tag{3.1}$$

where L is the inductance per unit length and C is the capacitance per unit length. The waves that use a transmission line have an E field to H field ratio that is equal to this impedance. When radiation is discussed in Chapter 6, the ratio of the E to H fields near a radiator is called the wave impedance.

The characteristic impedance that is used in most circuit board designs is $50\,\Omega$. Typically, this impedance is used even though there may be no attempt at using matching or terminating resistors. Lower impedances require higher current levels that can overload drivers. Higher characteristic impedances provide an opportunity for cross talk between traces. At higher logic speeds where lines must be terminated, the control of characteristic impedance becomes much more important. Circuit board manufacturers often provide added traces that can be used to test boards in production. Characteristic impedance provides a quick measure of plating thickness. Controlling this impedance controls the quality of the boards.

The characteristic impedance of various trace geometries can be found in the literature and on the Internet. There are field solvers that can handle more complex geometries. Note that there are many trace geometries that can provide the same characteristic impedance. It is good practice to work with a circuit board manufacturer before deciding on a trace geometry. There are many issues relating to cost and reliability that should be considered before a decision is made.

3.5 TRANSMISSION LINE FIELD PATTERNS

The B and H field patterns around parallel conductors carrying fields is shown in Figure 3.3a. The field pattern above the conducting plane is the same as the pattern in Figure 3.3b. Notice how the E field terminates on the conducting plane. This wave pattern is the same whether the voltage represents a logic signal or is used for dc power distribution. The E field lines are essentially perpendicular to the conductors. A very small component of the E field is located in the conductors to support any current flow. The characteristic impedance of the transmission line in Figure 3.3b is one half that of Figure 3.3a. The conducting plane in this figure is often called a ground plane.

A ground plane can be one of the conductors in a transmission line that carries energy. The fact that the current flows back to the energy source in the ground plane is just coincidental. This return current supports the magnetic field that is moving energy forward.

3.6 A TERMINATED TRANSMISSION LINE

When a wave reaches the end of a transmission line terminated in a resistor equal to its characteristic impedance, there is no reflection. The source voltage appears at the terminating resistor delayed by the transmission time. This wave action is shown in Figure 3.4. A more effective

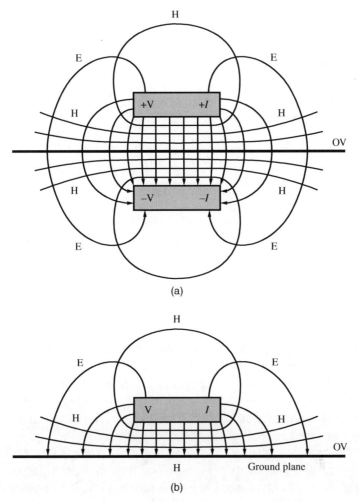

Figure 3.3   (a, b) The E and H field patterns around transmission lines

way to terminate a transmission line is discussed in Section 3.20, where a reflected wave returns a voltage equal to the source voltage and the current flow along the transmission line is set to zero.

In Figure 3.4, the transmission line is shown as two parallel conductors. After the switch closes, a wave travels to the terminating resistor and then all wave action stops. The voltage at the terminating resistor is equal to the source voltage. A trace over a ground plane, parallel traces or a length of coaxial cable that have the same characteristic impedance would perform in exactly the same way. In the figures that follow, the type of transmission line is not important and a single line representation is used. The voltages and impedances that are associated with

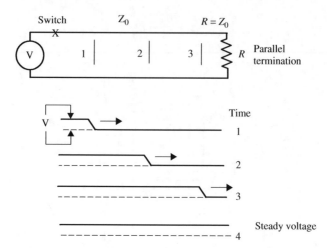

Figure 3.4 The wave associated with a transmission line terminated in its characteristic impedance

the transmission line are indicated on the line. The x along the line represents a switch. The infinity symbol indicates the line is long. The individual waves that reflect or transmit at the ends of the transmission line are shown on a set of additional time lines. Arrows indicate the direction of each wave after it is generated. The time selected for showing each wave position is somewhat arbitrary. The idea is to present the time history and direction of all the waves involved in the transmission.

Only one wave and all of its reflections are shown in Figure 3.4. Any number of waves can use a transmission line at the same time. Note that the energy in separate waves can be flowing in both directions at the same time. Energy placed in a transmission line is initially stored in the capacitance of the line. Eventually, this energy must be dissipated or radiated. Dissipation can be in resistances or in dielectrics. The energy must go somewhere. It cannot go back into the power source as that would be running "uphill."

3.7 THE UNTERMINATED TRANSMISSION LINE

When a wave reaches the end of an open (unterminated) line, energy cannot spill out into space. A reflected wave is generated that allows the energy to continue flowing from the voltage source. The two waves double the voltage on the line. This reflected wave cancels the current flow. When the reflected wave reaches the voltage source, the total voltage is double. For a step voltage, the source current goes to zero and a second

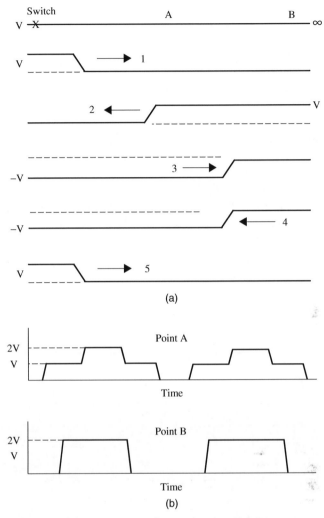

Figure 3.5   The voltage waveforms on an ideal open circuit transmission line for a step voltage

reflection sends wave energy back on to the line. If there were no losses, the voltage at the far end of the line would appear as a square wave of double voltage. If we view the voltage at the center of the transmission line, a staircase voltage would be apparent. The voltage at the source terminals would stay constant. In effect, the field energy supplied to the line during the round trip simply moves back and forth. In practice, the wave rapidly loses its character through losses and in a few cycles the voltage settles to a steady value. The waveform for reflections, assuming no losses or smearing, is shown in Figure 3.5.

In typical digital circuits, the terminations can be nonlinear and the reflections are not as simple as indicated here. Many logic gates that terminate a logic path appear as small capacitances. This tends to smear the leading edge. In this discussion, we will assume the leading edge retains its character.

3.8 A SHORT CIRCUIT TERMINATION

If the ideal transmission line is terminated in a short circuit, the first reflection must cancel the voltage. The current behind the reflected wave is doubled. The second reflection from the source adds new energy to the line. The current supplied from the source is now triple what it was originally. After a second round trip, the current is five times the initial value. This staircase of current flow continues until a fuse blows or the wires melt. This current pattern is shown in Figure 3.6.

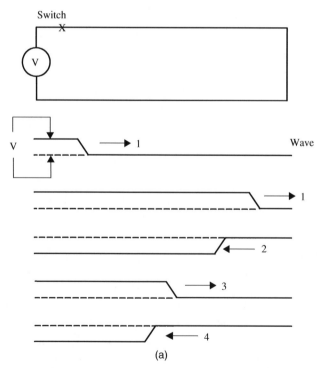

(a)

Figure 3.6 (a, b) The staircase current pattern for a short circuit termination of a transmission line

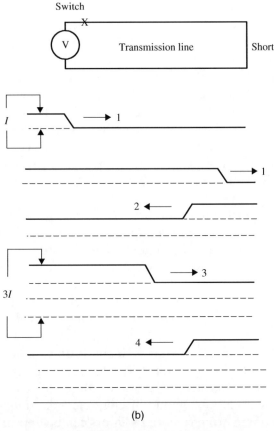

(b)

Figure 3.6   (*Continued*)

N.B.

A voltage source cannot sense the conditions at the end of a line until a reflected wave returns to the source. Time delays are not a part of circuit theory.

3.9 THE REAL WORLD

In the examples above, the transmitted signal is a step function typical of a logic transition. This transmission and reflection process would take place for any type of signal including sine waves, ramps, or noise. A ramp on the leading edge of a wave will show up in every reflection and transmission. Note that a general waveform can be formed by adding together a series of step waves that start at different times. If a transmission line works for one step wave and its ramp, it will work for any number of these waves in sequence. This means a transmission line will support waves of any shape.

Nature is not fussy. To her, all conductor pairs are transmission lines. Conductors that form pairs can include the earth, sheets of metal, conduits, shielded cables, open wires, building steel, telephone lines, conduit, and the power lines. These conductor pairs are irregular transmission lines with changing characteristic impedances. Many of these lines end in open or short circuits. Sine wave signals stay sinusoidal, but step waves (with a ramp) change character when they travel in complex structures.

Field energy from remote sources will couple to these odd transmission lines because less energy is required to use a conductor geometry than to travel in free space. This energy literally bounces around reflecting from all the irregularities. Light entering a room does the same thing. It reflects and gets absorbed at all the many surfaces. The light intensity at any one point in a room would be very difficult to calculate. Similarly, the field intensity in an arrangement of conductors is also very difficult to calculate.

Field coupling from local sources depends on whether the source is electric or magnetic in character. Close to changing high voltages, the coupling is usually capacitive in character. Near changing high currents the coupling is usually magnetic in character. At frequencies above 1 MHz, the coupling from remote radiators can be approximated using either field measure. At this point in our discussion, it is sufficient to say that almost all coupling is proportional to loop areas.

3.10 SINE WAVES VERSUS STEP VOLTAGES

Electrical engineers are very familiar with the analysis of linear circuits using sine waves.

This is the signal of choice because in a linear circuit, if the driving voltages are sine waves, all of the voltages and currents in the circuit will be sinusoids. Engineers in the power industry generate sine wave voltages so that a sinusoidal analysis makes sense. In the digital world, the signals involve step functions and time delays, and the best way to see what is happening is through examples in real time. The issues are more closely related to the flow of energy than they are to waveform analysis.

Logic signals have a frequency spectrum. The shorter the leading edge rise time, the higher the sine wave frequencies are that contributing to the wave. This is a factor when radiation or cross coupling is an issue. It is often useful to select a single frequency to do an analysis in order to get some idea of what the interference levels might be. See Section 6.4 for a further discussion of this method of analysis.

3.11 A BIT OF HISTORY

Before the digital age, analog circuits were built using point-to-point wiring, where a metal chassis provided a ground plane and shielding. Early digital circuitry used wire-wrap technology that was soon abandoned. This technology could not support the increasing logic speeds. Two-sided boards could not support a ground plane and the multilayer circuit board technology became a necessity. Over time, circuit density has increased, board costs have decreased, trace widths have decreased, and vias have become a way of life. More and more logic and memory have been placed in semiconductor components and operating speeds have increased dramatically. Many isolation techniques have been developed that include rf and optical links, smaller connectors, and better board materials. The one thing that has not changed is the basic physics. Radiation that results from shorter rise times has become a bigger issue. Multipin connections to components are forcing additional board layers just to connect to all the pins. This in turn makes it more difficult to control the transmission line character for logic lines near the integrated circuits.

3.12 IDEAL CONDITIONS

Nothing happens in zero time. In digital logic, picoseconds can be a consideration and for humans that comes very close to being a zero time. One of the problems with discussing the way energy moves between components is that ideal switches must be used. If the logic operates at 1 GHz, a rise time of 10 ps is important. The problem is further complicated when we recognize that the switch we use must be much smaller than the circuit so that it is not a part of the circuit. We must use ideal switches in our discussions as this is the only way we can gain insight into circuit operations.

We will use ideal switches, ideal terminations, and in many cases ignore lead length and lead inductance. In our analysis, we will assume a zero impedance voltage source at all frequencies. In many of our discussions, a 16th inch of trace is a transmission line. Another more serious limitation is the real character of termination resistors. A typical resistor has at least 1 pF of shunt parasitic capacitance. When mounted on a circuit board, this can be 3 or 4 pF. At 1 GHz, this is a reactance of less than 50 Ω. This does not consider any series inductance or any nonlinearities. A resistor can also be thought of as a short segment of lossy transmission line. This means that terminating a fast logic transmission line is only going to be an approximation.

3.13 REFLECTION AND TRANSMISSION COEFFICIENTS

Before we examine energy flow on a circuit board, we need two important equations. The first is the reflection coefficient. Consider cascading transmission lines having different characteristic impedances. This is exactly the same problem as terminating a transmission line in a resistor other than its characteristic impedance. Obviously, if the termination is a match, the energy flows through the transition without a reflection. We already know that if the termination is an open circuit (infinite impedance), the reflected wave equals the forward wave and the voltage doubles. If the terminating impedance is $0\,\Omega$, the reflected wave is the negative of the forward wave and the resulting voltage is zero. The reflection coefficient is

$$\rho = \frac{(Z_1 - Z_0)}{(Z_1 + Z_0)} \tag{3.2}$$

where the wave travels from Z_0 to Z_1. If $Z_1 = Z_0$, then $\rho = 0$. If Z_0 is very large, then $\rho = -1$. If the initial wave is V_0, the reflected wave is ρV_0. If the reflection occurs at a voltage source, then Z_1 is considered to be $0\,\Omega$. If the reflection occurs at an open circuit (such as a transistor gate), then Z_1 is considered infinite.

The transmission coefficient at an impedance transition is

$$\tau = \frac{2Z_1}{(Z_1 + Z_0)}. \tag{3.3}$$

If $Z_1 = Z_0$, then $\tau = 1$. If Z_0 is large, then $\tau = 0$. If the initial wave is V_0, the transmitted wave is τV_0.

On a complex structure, the voltage level on any transmission line segment is the sum of the initial condition and all the transmitted and reflected wave voltages that have passed that point up to the time t. The velocity of a wave front is given in Equation 3.4, where c is the speed of light and ε_R is the relative dielectric constant. This equation can be used to locate the position of each wave front as reflected and transmitted waves are generated.

$$v = \frac{c}{\sqrt{\varepsilon_R}}. \tag{3.4}$$

For glass epoxy, the relative dielectric constant is approximately 4. On a circuit board, a wave travels about 15 cm/ns.

3.14 TAKING ENERGY FROM AN IDEAL ENERGY SOURCE

Consider a 50-Ω transmission line that is 10 cm long connected to an ideal 5-V source. At time $t = 0$, connect a 50-Ω load resistor to the end of the transmission line. See Figure 3.7 for the circuit and the voltage waveforms.

When the switch closes, the current in the forward and reflected waves must add up to zero. This can only happen if a wave W_1 of −2.5 V travels toward the voltage source and a wave W_0 of +2.5 V moves toward the resistor. The voltage on the transmission line is now the sum of the initial voltage and W_1 or 2.5 V. When W_1 reaches the 5 V source, there is a reflection from a zero impedance point. The forward wave W_2 is of opposite sign of W_1 or +2.5 V. The sum of W_1 and W_2 is 5 V. When this 2.5 V wave reaches the resistor, the total voltage is 5 V and wave action stops.

In Figure 3.8, the transmission line does not match the load impedance and waves must make many round trips before the voltage is near the correct level. This mismatch results in a delay that can be troublesome. In the examples above, the step character to the voltage at the load resistor would be difficult to observe. The waveform at the resistor would appear smooth and exponential in shape as if this was a delay caused by inductance. The short transmission line actually has inductance, but it also has capacitance. As pointed out earlier, both the magnetic and electric fields are needed to move energy. The amount of energy that can be carried by a wave is determined by the voltage level and the characteristic impedance of the line.

3.15 A CAPACITOR AS A TRANSMISSION LINE

A capacitor is a two-terminal device. As discussed earlier, energy cannot be removed and resupplied at the same time. The relative dielectric

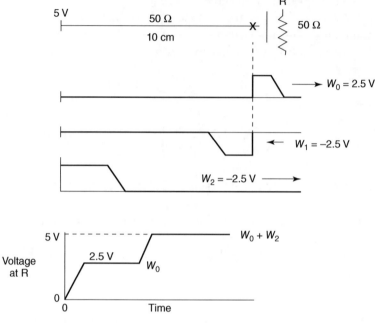

Figure 3.7   Waves that supply energy to a resistor over an impedance matching transmission line

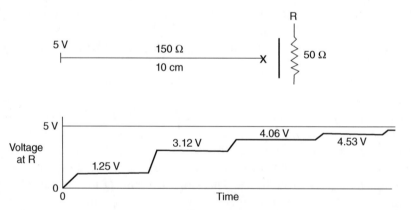

Figure 3.8   The voltage at a termination resistor when there is a mismatch of impedances

constant used in decoupling capacitors can exceed 10,000. This means that the velocity of a wave in the capacitor is a hundred times slower than in air. A capacitor that is only 0.2 cm long can be electrically 20 cm long. The characteristic impedance of this line can be as low as 1 Ω.

The basic problem with applying decoupling capacitors to supply energy to a component is that they function as transmission line stubs when placed along a section of 50-ω trace. The distance from the stub to the integrated circuit can be a centimeter. At the same voltage, the maximum energy that can be carried by a wave on a 50-Ω line is 1/50 the energy per wave on a 1-Ω line. This means that many round trips on a 50-Ω line are required to move useful energy out of a decoupling capacitor. Again this limitation is often blamed on series inductance. The solution to this problem is to keep trace length to an absolute minimum. The single port aspect of a capacitor is still a problem.

In applications where many parallel high-speed drivers are used, the energy demand can be significant. In these applications, the practical solution is to provide many parallel decoupling capacitors mounted at the periphery of the component. A single large capacitor value is not usually effective. Parallel stubs allow energy to be supplied from several sources at one time.

In some applications, rf energy must be contained inside a metal housing. Any unshielded lead leaving the enclosure would allow energy to radiate. This problem is usually resolved by using a "feed-through" capacitor. This comes close to being a four-terminal (two-port) capacitor. The capacitor is located in a short section of threaded coax that can filter a lead that crosses a conducting boundary.

3.16 DECOUPLING CAPACITORS AND NATURAL FREQUENCIES

Capacitors are often manufactured using stacked layers of dielectrics and conductors. The conductive layers are then bonded together to form the two terminals of the capacitor. The bonding geometry determines the path fields must take to enter the dielectric. If the field path is blocked in any way, the bulge in field adds series inductance. This is another way of saying that the characteristic impedance of the transmission path has been compromised.

Surface-mounted capacitors are used in a typical board design. They are amenable to automatic assembly methods and they do not have lead inductance. It is common practice to view the natural frequency of capacitors rather than their transmission line character. Their resonant frequency is where the apparent series inductive reactance equals the reactance of the capacitance. Capacitors with small capacitance values have higher natural frequencies. One approach in design is to populate the board with several capacitor sizes so that energy is available over a wide frequency spectrum. One rule that is followed is to locate the smaller valued capacitors near the active components requiring decoupling.

A wide range of decoupling capacitors is available on the market. Ball-grid array capacitors perform the best because they can be mounted very close to the point of demand. The natural frequency of a typical 0.01 μF decoupling capacitor is 65 MHz and for a 0.001 μF capacitor the natural frequency increases to 112 MHz.

Above 100 MHz, the character of decoupling capacitors begins to change. As pointed out earlier, the electrical length of a capacitor reinforces the fact that at these frequencies, this component is transitioning to a transmission line. The resonance phenomenon that is measured assumes a simple inductance and capacitance. In fact, the actual conductor geometry is closer to a combination of lumped and distributed parameters. This is a situation where the semantics of the past may not be appropriate. The real issue rests not in natural frequencies but in how to obtain energy in short periods of time.

N.B.

Manufacturers of integrated circuits will often specify the size, type, and number of decoupling capacitors needed for a component to perform.

The traces on circuit boards carry and store some energy. Often the initial energy demands of a circuit come from these trace geometries rather than from decoupling capacitors. These energy sources are located close to the point of demand. These sources of energy are not shown on any schematic, but their presence can be needed for the circuit to function. These parasitic energy sources include logic traces connected to logic "one", traces connected to decoupling capacitors and the connections to the ground/power planes.

3.17 PRINTED CIRCUIT BOARDS

The transition from wire-wrap to printed circuit wiring was needed to reduce manufacturing costs and improve circuit performance. This manufacturing art served both analog design and the slower logic of the day. At first, power and ground conductors were just leads. As the logic became faster and more complex, the need for ground planes became obvious.

It did not take long and multilayer board technology was offered to board designers.

Multilayer printed circuit boards are manufactured by bonding together layers of copper and glass epoxy. The stack of materials is called a *layup*. On a four-layer board, the core or center of the board is a sheet of epoxy laminated on both sides with copper. The copper is used as a conducting plane or etched to form traces. Holes are drilled and plated. The outer copper layers of the board are drilled, etched, and plated. Then they are bonded to the core using layers of partially cured epoxy called *prepreg*. The *layup* is cured under heat and pressure. The assembly is then drilled and plated as required. Many different procedures and techniques are used depending on the number of boards being manufactured, the number of layers, component density, and application.

Structural integrity and the ability to handle heat means that very thin board designs are more difficult to build. Boards that are 0.062 in. thick have become fairly standard. In many digital designs, transmission lines are short and do not require termination. It is an accepted practice that the characteristic impedance of traces be controlled. This control involves trace width, thickness, and spacing. One approach is to dedicate two layers to ground (logic 0) and to mix logic with ground and power plane islands on the two remaining layers.

A ground plane can be broken up into islands as long as all logic traces that use the island as a return path do not leave the island area. If a trace crosses over a split ground plane and there is no controlled return logic path, there will be reflections and there will be radiation. A power plane functions just like a ground plane except that it is associated with a dc voltage. The dc level can be thought of as an earlier wave that is not changing value. Remember that any number of waves can use a transmission line at the same time.

3.18 TWO-LAYER LOGIC BOARDS

Many analog and unclocked logic boards can be built using a two-layer board. The leads should be carefully routed to limit loop areas for all signal paths. It is also practical to build some clocked logic circuit boards that meet radiation and susceptibility standards using a two-layer board. It takes more traces than in a four-level design, but then again large surface areas for ground and power planes are not used. In this approach, every trace carrying logic is placed between a ground and power trace. This triplet transmission line goes in the x-direction on one side of the board and in the *y*-direction on the other side of the board. The triplet line changes direction by passing through a triplet of vias. On a regular pattern, the ground and power traces on the top of the board are connected using vias to the ground and power traces on the bottom side of the board. This forms a grid that looks very much like two conducting planes. The field for every signal is closely contained. This triplet pattern supplies power to every logic component. This geometry provides field control and reduces board costs in many cases.

3.19 VIAS

DEFINITION

A "via" is a conducting path in a printed circuit board between layers. The path is usually a plated through-hole in the board. Note that logic fields cannot pass through a plated hole. Consider that the E field lines must be perpendicular to the conducting surface, and the problem becomes clear.

FACT

Fields carrying logic cannot pass through conductors.

DEFINITION

A "blind via" is visible from the outer surface, but the hole is blocked by a conductor on an inner layer.

DEFINITION

A "buried via" is located between inner layers and is not visible from outside the board. This via is also called an embedded via.

N.B.

A through-hole is used to mount components. It can also serve as a via.

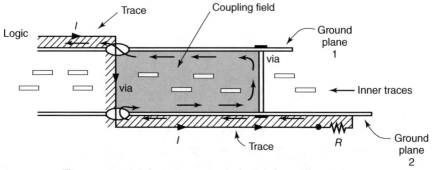

The trace current "I" crosses to a new layer using a via, the return current must take a complex path through holes in the ground planes because high-frequency current cannot flow through a conducting plane. The resulting field patern is complex and causes cross coupling.

Figure 3.9 A via geometry

When a via connects a logic or signal trace from one layer on a circuit board to another, the return path must be on a separate ground via. In high-speed logic, the two vias should form a short section of transmission line that carries logic or signal. If the characteristic impedance is not controlled in this transition, reflections and cross talk can limit performance and add to the radiation from the board. The fact that the ground planes are already tied together is not sufficient. Figure 3.9 shows a pair of transmission line vias. Note that the field travels through holes in the ground plane. The pads on the ends of the vias provide these openings.

N.B.

Logic fields cannot cross through a conducting plane. To get to the other side of a conducting plane, fields may have to go around to the edge of a board. This forces a reflection and is a source of radiation.

It is a good idea on multilayer boards to draw the field path when vias are involved. This involves the outgoing and return current paths. The E field in this path crosses between the forward and return conductors. The energy flow direction is perpendicular to the E field. If the path is circuitous (no pun intended) or if the field pattern bulges, there is a problem.

> **N.B.**
>
> Fields cannot cross through a conducting surface.

Fields can cross through a conductor only where there is an opening in the conductor and there is a nearby trace to direct the field flow.

A simple via that connects parallel ground planes may only connect the inner surfaces. To connect the outer surfaces of parallel ground planes, a via must terminate on pads that provide openings in the planes. These openings must be associated with every logic trace that transitions between conducting planes. Just having openings is not sufficient.

When a trace uses a ground plane as a return conductor in one area and a power plane in another, the field path must involve a decoupling capacitor. The current path for the transmission line can be complicated. Since there are fewer capacitors than traces on a circuit board, their location near components takes priority. This practice is therefore not recommended.

3.20 THE TERMINATION OF TRANSMISSION LINES

When a positive wave on a transmission line reaches an open termination at a logic gate, the reflected wave can double the voltage. This may damage the component or initiate a logic error. Terminating the line in its characteristic impedance eliminates this problem as there is no reflection. This may not be a viable solution as the logic level might have to put energy into the termination for long periods of time. The preferred way to terminate a transmission line for fast logic is to locate the terminating resistor at the logic source. This is called a series termination as opposed to a shunt termination. A 50-Ω source resistor and a 50-Ω transmission line will reduce the wave amplitude to half value. At the open end of the line, the doubling action brings the logic level up to full value. There is no dissipation at the open end of the line. When the reflected voltage reaches the source, the source voltage is equal to the logic level, the current flow is zero, and all wave action stops. There is no further dissipation. The only problem is that the elapsed time before the energy stops moving is one round trip of the wave. When the next wave returns the logic level to 0 V, the energy stored in the line capacitance will be dissipated mainly in the series termination. If a terminating resistor is not used, most of the energy stored in the line capacitance will be

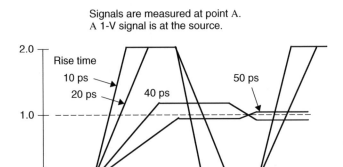

Figure 3.10   Rise times and the reflections from an unterminated transmission line

dissipated in the driver. There is no way to avoid the fact that this field energy must be dissipated.

N.B.

To be effective, a series terminating resistor must be located very near the driver.

In many logic designs, the line length is short enough that no reflections occur. In this case, terminations are not needed. The nature of the problem is shown clearly in Figure 3.10, where the reflection does not occur if the rise time is long enough.

The velocity of a wave on a circuit board transmission line is usually half the speed of light. The equation for velocity in meters per second is given by

$$v = \frac{1}{\sqrt{LC}} \tag{3.5}$$

where L is the distributed inductance in henries per meter and C is the distributed capacitance in farads per meter. Since capacitance is inversely proportional to the dielectric constant, the velocity of a wave is inversely proportional to the square root of the dielectric constant. The dielectric constant of epoxy is about 4. For sine waves at 1 GHz, this figure falls to 3.5. Typically, the velocity along a circuit trace is about 6 in./ns. The distance a signal travels during a logic rise time provides a good reference point. If a transmission line is shorter than one-fourth of this distance, then transmission line terminations are usually not required. For example, consider a 100 MHz clock with a rise time of 2 ns. If the traces are less than 3 in. long, there is little need to terminate the lines. At 1 GHz, this distance reduces to 0.3 in. Consider a transmission line that is 20 ps long. The waveforms at an open circuit for the doubling action shown in Figure 3.10 can be scaled. If the rise time is doubled, then the length of unterminated line can be doubled.

3.21 ENERGY IN THE GROUND/POWER PLANE CAPACITANCE

The ground and power plane does provide some decoupling energy. When a logic component is connected to the ground and power planes, the energy in the capacitance is not immediately available. A typical connection to this capacitance is a short transmission line. The planes form a conductor geometry that is a tapered transmission line, where the characteristic impedance falls off proportional to the radial distance. For the wave that propagates outward, the reflection process is continuous. The characteristic impedance falls off linearly with distance and a low impedance is not immediately available. It turns out that a high dielectric material does not help as it slows the progress of the wave. The only parameter that shortens the time it takes to get energy is the thickness of the dielectric. Manufacturing a board with a small enough spacing between layers is not practical. Decoupling capacitors are far more effective than a ground/power plane in supplying energy.

The ground/power planes are connected to both decoupling capacitors and active circuits. Every logic transition places a small demand for energy stored in this capacitance. This wave action reflects and transmits at all component connections, vias, gaps, pads, and at the board edges. This is an example of where countless numbers of waves are in motion at any one time over the entire board. This amounts to background noise for every logic line on the board.

Circuit traces can be placed on layers between the power and ground planes. The fields associated with these traces are tightly confined so there is little chance of radiation. For a centered trace, there are two field paths. A single (source) terminating resistor can be used. A mix of transmission line types can pose a problem if two field patterns must merge back into one termination.

N.B.

A power plane functions just like a ground plane except that it is associated with a dc voltage.

3.22 POYNTING'S VECTOR

An important concept in electromagnetism theory is Poynting's vector. The power density of a field moving past a point in space is equal to the vector cross product of the E and H field vectors. We can cheat a bit and say power per unit area at a point is simply E times H. Poynting's vector has intensity and direction at all points in space. The total power crossing a surface is the integral of Poynting's vector over that surface area. Figure 3.11 shows Poynting's vector P for two conductors carrying power. The vectors E, H, and P are always at right angles to each other. The E field has units of volts/meter, the H field has units of amperes/meter, and the product has units of watts per meter squared.

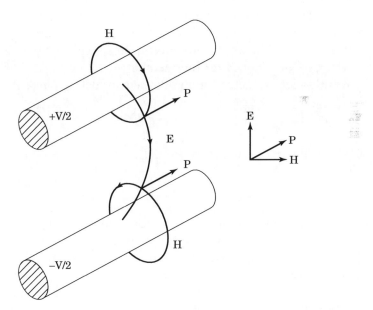

Figure 3.11   Poynting's vector for parallel conductors carrying power

The vector applies to fields moving between conductors as well as to fields in space that are radiating.

3.23 SKIN EFFECT

In the first chapter, the electric field lines terminated on surface electrons. In the second chapter, dc current flow generated a magnetic field, where it was assumed the current used the entire conductor in a uniform manner. For alternating currents, the magnetic field in the conductor limits the current flow as a function of depth. This lack of penetration is called skin effect. In the power industry, this lack of penetration at 60 Hz means the copper at the core of a large diameter conductor is not fully utilized. For long transmission lines where support towers are used, it makes economic sense to use steel in the core of the conductors. This allows for fewer towers to support the conductors. For analog circuits, copper is such an excellent conductor that a lack of penetration rarely causes a problem. For most digital circuits, the currents stay on the surface of conductors. Fortunately, on most circuit boards, the path length is short and any increased resistance is not a serious problem.

The equation for skin effect on round or rectangular conductors is complex. The approximation that is generally used is that of field penetration when a plane electromagnetic wave reflects off of a plane conducting surface. The depth of penetration for this ideal case is used to approximate the depth for other geometries.

The attenuation factor for an infinite conducting plane is

$$A = e^{-\alpha h} \tag{3.6}$$

where h is the depth of penetration. The term α is equal to

$$\alpha = (\pi \mu_0 \alpha f)^{1/2} \tag{3.7}$$

where μ is the permeability, σ is the conductivity, and f is the frequency in hertz. The relative permeability of copper is 1 and the permeability μ_0 of free space is $4\pi \cdot 10^{-7}$ H/m. The conductivity of copper is $0.580 \cdot 10^8$ A/(V m). At 1 MHz, $\alpha = 15.13$/mm.

When $h = 1/\alpha$, the attenuation factor is $1/\varepsilon$. Expressed in decibels, the attenuation factor is -8.68 dB. This depth is called a skin depth. For copper at 1 MHz, the skin depth is 0.066 mm. Skin depth is inversely proportional to the square root of frequency. The skin depth of copper at 100 MHz is 1.51 mm. At two skin depths, the attenuation is 17.36 dB.

At 60 Hz, the skin depth of copper is 0.855 cm. This is an approximate value for round or rectangular conductors.

Skin effect limits the penetration of logic current in a conducting plane. One ounce copper plating means one ounce of copper per square foot of surface area. The thickness of 2-oz. copper is about 0.3 mm. The ohms per square for this thickness of copper at 10 MHz is 390 μΩ. The skin depth at 10 MHz is only 0.02 mm. When traces are formed by etching, the layer thickness controls the trace thickness that in turn controls the characteristic impedance of the transmission path.

3.24 MEASUREMENT PROBLEMS: GROUND BOUNCE

The loop areas formed by an oscilloscope probe tip and the probe shield are large compared to the loop areas on the circuit board. This means that observed signals may include fields from nearby circuitry. Moving the reference connection to a point nearer the signal point can reduce the loop area at the probe tip. Reducing the loop area will limit some of the coupling.

The radiated field near the board may use the probe braid as a conductor and allow energy to leave the circuit. The probe here is functioning as an antenna. Current flowing on this braid can cross couple some field into the probe center creating an observable signal. To test for this coupling, the probe common and tip can be tied together and disconnected from the circuit board. A better probe shield may be required to eliminate this coupling.

Another test that is often performed is to connect an oscilloscope between two ground points on a circuit board. If a voltage is detected, it is often blamed on an inadequate "ground." The ohms per square for a copper ground plane even at 100 MHz are only a few milliohms. A volt of signal associated with the ground plane would imply a 1000 A of current flow. The only way to explain this voltage is to recognize that there must be a changing electromagnetic field in the area near the board. Creating a loop with the same area and using it to sense the field can verify this. The loop should be oriented the same way as the oscilloscope probe and it should be left floating.

When ground potential differences are observed, the effect is called "Ground Bounce." The term is unfortunate because it suggests ground current. To reduce this effect, the loops that are creating the field must be made smaller.

A measure of the voltage drop across a length of trace will usually be very misleading. The voltage that is measured is from fields that are not properly confined by the traces and the ground plane.

The signal patterns along a transmission path on a circuit board are usually not observable. The sharp patterns shown in the figures are not often displayed. The waves are shown as line segments so that the character of the waves is emphasized. In practice, traces are usually covered by a layer of dielectric to prevent moisture from penetrating the board. Moisture has a high dielectric constant that can modify the characteristic impedance of the line. Further, the presence of a probe can introduce a stub that can alter the very signal that is to be observed.

Cross talk between transmission lines can introduce errors. This subject is treated in the author's book titled *Digital Circuit Boards – Mach 1 GHz*, Sections 3.8, 3.9–3.11. This book is also published by John Wiley & Sons in 2012.

3.25 BALANCED TRANSMISSION

Logic signals must often be carried between circuit boards, where a continuous ground plane is not available. As we have seen, a grounding conductor will not eliminate potential differences between ground planes. An added conductor in a cable cannot eliminate fields. The result is that an interfering voltage can be added to every logic transition in a connecting cable. To handle this problem, designers often use a balanced signal and a balanced logic receiver. A second transmission line that carries inverted logic is used to create a balanced signal. If line one is $+3\,V$, then line two is $0\,V$. If line one is $0\,V$, then line two is $+3\,V$. This pair of signals is balanced as their sum is a constant. The potential difference between circuit grounds adds the same error signal to both lines. The receiving logic subtracts out this error signal and generates a single logic signal referenced to the receiving ground. In analog parlance, the ground difference of potential is called a common-mode signal as opposed to a normal-mode signal. In digital parlance, the error signal is called an odd-mode signal as opposed to an even-mode signal.

The treatment of an even-mode signal on the receiving board is shown in Figure 3.12.

Assume the characteristic impedance of each arriving logic line is $50\,\Omega$ with respect to the common shield. When the cable is terminated on the receiving board, each line is matched to the 50-ω characteristic impedance of traces on the board. The cable shield is connected to the ground plane. No terminating resistors should be used if there is trace length on the board.[1] The two signal traces are kept separate

---

[1] If the incoming transmission line has a characteristic impedance of $100\,\Omega$, then any attempt to match this impedance where the line terminates will doubly terminate the

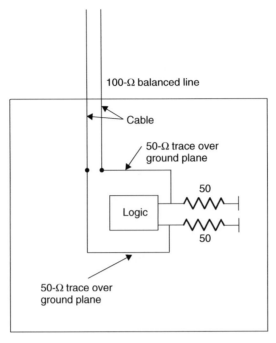

100-Ω balanced line

Cable

50-Ω trace over
ground plane

50

Logic

50

50-Ω trace over
ground plane

Figure 3.12   The termination of a balanced transmission line on a circuit board

on the board until they reach the receiving component. The two path lengths should be equal to avoid a difference in arrival time. Each logic line should be terminated in 50 Ω at the receiving logic. Remember, termination resistors are required as the signal path length includes the external cable length. A long transmission line requires termination to avoid voltage doubling. Note that if one of the balanced lines should couple to an error signal, then the resulting error cannot be removed by the receiving circuitry. The two signal paths should be kept apart so that the two transmission fields do not share the same physical space.

3.26 RIBBON CABLE AND CONNECTORS

Ribbon cables are insulated, parallel, evenly spaced conductors. The cable provides a very orderly way to carry a group of conductors

line. If the board has a ground plane, then the connection to the board will automatically provide a 50-Ω matching impedance for each line. If the lines are closely spaced on the board, then there will be a mismatch in impedances. Of course, if the receiving hardware is located at the transmission line connection point, then the two 50-Ω resistors should be located at this point. One 100-Ω terminating resistor is not recommended as common-mode signals can double at an open termination.

between two points. Ribbon cables can be soldered into place or terminated in connectors. Once conductors leave a circuit board, they are no longer closely spaced or close to a ground plane. The conductor length and increased spacing adds to the loop areas, and this increases the chances of radiation, reflections and susceptibility.

It is good practice to provide more than one ground conductor in a ribbon cable carrying logic. If every third conductor is a ground conductor, then every logic conductor has a nearby return path. This scheme will not be effective unless the ground conductors are individually carried through the mating connectors and are terminated on ground planes at both ends of the cable.

Ribbon cables are available with a copper backing on one or both sides of the cable (Microstrip or Stripline). This backing can serve as a ground plane provided the plane is properly terminated at the ends of the cable. This copper should not be considered a shield but an extension of the ground plane. A single connection to this sheet of copper at either end of the ribbon run will negate the value of the plane. Multiple connections are required as this conductor is not a shield. It is a ground plane.

Ribbon cables frequently cross open areas. This increases the chances of common-mode coupling. To limit this coupling, the cables should be routed along conducting surfaces. Excess ribbon cable should not be coiled as this also increases the opportunity for common-mode coupling.

3.27 INTERFACING ANALOG AND DIGITAL CIRCUITS

It is often necessary to interface analog and digital circuitry. At digital speeds, the differential approaches used in analog coupling are hardly appropriate. There are several schools of thought that include separate grounding schemes and even separate boards. Somewhere the analog ground must come in contact with the digital ground. The problem is usually related to an analog-to-digital (A/D) converter. If 14-bit accuracy is required, the error for a 10-V full-scale signal is only 0.5 mV. If two grounds are involved, the noise voltage that is sensed can easily exceed 10 times this figure. If the two grounds are connected together by a single conductor, the loop area that is formed is great enough that the ground potential difference will still be a problem.

The best solution for interfacing analog and digital functions is to use a common unbroken ground plane. It is important to make sure that the fields associated with analog and digital functions do not share the same physical space. The following rules apply:

1. Analog circuit components must not be intermixed with digital circuit components.
2. Make sure that the analog fields do not share the same physical space as the digital fields.
3. The pin assignments in connectors should separate function so that the analog fields and digital fields use a different space.
4. The A/D converter should have an internal forward referencing amplifier.
5. Orient the A/D converter to limit field coupling.
6. The analog and digital circuits should pull energy from different decoupling capacitors.

CHAPTER 4

Analog Circuits

OVERVIEW

This chapter treats the general problem of analog instrumentation. The signals of interest are often generated while testing functioning hardware. Tests can take place over time, in a harsh environment, during an explosion, during a flight, or in a collision. The signals of interest usually have dc content and can be generated from floating, grounded, balanced, or unbalanced transducers. These transducers may require external balancing, calibration, or excitation. Accuracy is an important consideration. Where data must be sampled, the signals may require filtering to avoid aliasing errors. The general two-ground system is examined. Protecting signals using guard shields, transformer shields, and cable shields is described. The use of feedback and tests for stability in circuit design is considered. Strain-gauge configurations, thermocouple grounding, and charge amplifiers are discussed.

4.1 INTRODUCTION

The expression "analog circuit" can mean any electronics that is "nondigital." An rf-transmitting circuit can be considered an analog circuit. This chapter is devoted to analog circuits that operate below 100 kHz. The techniques that are described can be applied to audio amplifiers, medical electronics, and power supplies as well as instrumentation.

The availability of integrated circuits has simplified many aspects of analog circuit design. Instrumentation must often handle long signal lines, reject ground potential differences, and maintain circuit stability.

Grounding and Shielding: Circuits and Interference, Sixth Edition. Ralph Morrison.
© 2016 John Wiley & Sons, Inc. Published 2016 by John Wiley & Sons, Inc.

The general problem of analog design is called signal conditioning, which includes gain, filtering, offsets, bridge balancing, common-mode rejection, transducer excitation, and calibration. In large permanent installations, once a signal has sufficient resolution and the bandwidth has been controlled, the signal can be digitized and transmitted over an rf link to a computer, where further conditioning can be provided. This chapter treats the problems of conditioning signals before they are sampled and recorded.

This book places emphasis on electromagnetic fields. These fields are the basis of all electrical behavior. In analog work, it is useful to describe problems in terms of circuit behavior. Fields will still be discussed when it provides needed insight.

4.2 INSTRUMENTATION

Measurements of temperature, strain, stress, position, and vibration are an important part of many development efforts. Aircraft landing gears, missile housings, helicopter rotors, and turbine engines are examples of structures that undergo extensive testing both in development and production. Transducers associated with these measurements are usually mounted directly on a test structure or vehicle. Some of these transducers require external excitation while others are self-generating. Some transducer types are electrically bonded to the structure under test and others are electrically floating. The signals generated by these transducers are usually in the millivolt range. These signals must be amplified, conditioned, and then recorded for later analysis. The signal conditioning can be located on the structure under test, or it can be located at a separate location.

There are many ways for an analog signal to be compromised. It makes no sense to use digital signal filters if the signal has already overloaded the circuitry. If shields are not terminated correctly, software cannot remove the resulting interference. If common-mode signals are not rejected, significant errors are the result. If the signal has not been filtered, then sampling can introduce aliasing errors that can make the data useless. All of these problems are discussed in this chapter.

N.B.

If the answer is unknown, it can be very difficult to verify that the measurement is valid. For example, signals that overload an input stage can produce noise that may look like signal.

There are semantics problems in discussing analog electronics. Here are a few definitions that will help in understanding the discussions that follow.

1. *Reference conductor.* Any conductor used as the zero of voltage. In a power supply with voltages of 0, +15, −15, and −5V, the conductor labeled 0 V is the reference conductor. If a signal is measured with respect to a conductor called ground, it becomes the reference signal conductor. In an analog circuit, there may be several reference conductors in one amplifier. For example, input common, output common, and excitation common are all different reference conductors and they are all a part of one instrument. A second instrument will have its own reference conductors.

2. *Signal common.* A signal reference conductor

3. *Signal ground.* A signal reference conductor

4. *Balanced signal(s).* Two signals measured with respect to a reference conductor whose sum is always zero. For example, a centertapped transformer produces a balanced signal. The centertap can be called the reference conductor and the voltages on the transformer terminals $+V_{sig}$ and $-V_{sig}$ are a balanced signal. The signal generated by active arms in a Wheatstone bridge can be a balanced signal.

5. *An unbalanced signal.* A single signal voltage measured with respect to a reference conductor. A single-ended signal.

6. *Common-mode voltage.* The average interfering voltage on a group of signal conductors measured with respect to a reference conductor. This is usually a ground potential difference. In telephony, a common-mode signal is called a transverse-mode signal. In logic, a common-mode signal is an odd signal.

7. *Normal-mode signal.* The signal of interest. In telephony, the normal-mode signal is called the transverse-mode signal. In logic, a logic signal is called an even signal.

8. *Differential signal.* The voltage difference of interest.

9. *Difference signal.* The voltage difference of interest.

10. *Instrumentation amplifier.* A general-purpose differential amplifier with bandwidth from dc to perhaps 100 kHz and variable gains from 1 to 5000. This instrument might provide transducer excitation, filtering, and signal conditioning. The material in this chapter stresses handling analog signals.

4.3 HISTORY

The first signal amplifiers were designed using vacuum tubes. The dc operating voltages often exceeded 200 V and filament current had to be supplied to each tube. Since vacuum tubes are not available as complimentary devices (e.g., NPN and PNP transistors), early instrumentation amplifiers were ac amplifiers. Low-frequency signals were often mechanically or electrically modulated and amplified as a carrier. Strain gauges were excited directly with a carrier signal. The amplification of carrier signals allowed the use of transformers that provided isolation between input and output circuits. After amplification, the signals were demodulated and filtered. The end result was amplification with a limited bandwidth. Bandwidths using carrier techniques rarely exceeded 1 kHz. In this approach, cross talk between channels was very difficult to control.

In telephony, dc amplification is not needed and transformers can be used to provide isolation and to convert single-ended audio signals to balanced signals and vice versa. Balanced telephone signals can be transported over long distances on wires and the effects of common-mode interference are canceled. In audio work, transformers can eliminate circuit common connections (ground loops). Feedback around transformers is usually not practical.

When semiconductors were introduced, it became possible to amplify signals with a dc component using well-balanced matched pairs of transistors. Techniques evolved that made signal isolation possible without the use of transformers. Today, designs are available that handle microvolt to 100-V signals with bandwidths from dc to many megahertz. Millivolt signals can be amplified in the presence of 300-V common-mode voltages. Conditioning various signals with dc content in the presence of interference is the subject of this chapter.

Newcomers to instrumentation often take a simplistic view of the measurement problem. The problem is much more complicated than simply providing gain. It takes experience to bring the full range of difficulties into focus. The story involves circuit design, specifications, and application. The material that is presented here is for the instrument designer as well as for the user. It is important that each side understands the issues. Getting it right is the issue.

4.4 THE BASIC SHIELD ENCLOSURE

We will start with some very elementary ideas. The simple amplifier circuit shown in Figure 4.1a has potentials labeled as follows: the input

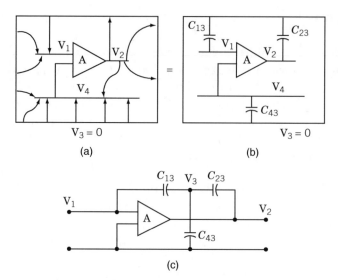

Figure 4.1   Parasitic capacitances in a simple circuit. (a) Field lines in a circuit, (b) mutual capacitance diagram, and (c) circuit representation

lead is V_1, the output lead is V_2, and a signal common or reference conductor is labeled V_4. The conducting enclosure labeled V_3 is floating. It is convenient to call the enclosure the reference conductor at 0 V. Every conductor pair has a mutual capacitance, which we label C_{13}, C_{23}, C_{43}, and so on. When these capacitances are drawn out as a circuit in Figure 4.1b, it is immediately apparent that if the circuit is an amplifier there is feedback from the output to the input. The circuit is shown in Figure 4.1c.

It is common practice in analog design to connect the enclosure to circuit common. This enclosure connection is shown in Figure 4.2. When this connection is made, the feedback is removed and the enclosure no longer couples signals into the feedback structure. The conductive enclosure is called a shield. Connecting the signal common to the conductive enclosure is called "Grounding the Shield." This "grounding" usually removed "hum" from the circuit.

Most practical circuits must provide connections to external points. To see the effect of making a single external connection, open the conductive enclosure and connect the input circuit common to an external ground. This ground can be any structure, earth or hardware common. Figure 4.3a shows this grounded connection surrounded by an extension of the enclosure. This extension over the input conductor pair is called a cable shield. A problem can be caused by an incorrect location of the connection between the cable shield and the enclosure. In

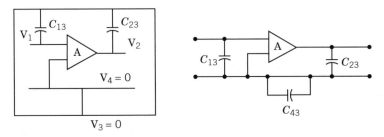

Figure 4.2   Grounding the shield to limit feedback

Figure 4.3a, the electromagnetic field in the area induces a voltage in the loop ① ② ③ ④ ① and the resulting current flows in conductor ① ②. If this conductor is the signal common, this lead might have a resistance R of 1 Ω. In this case, every milliampere of coupled current would develop a millivolt of interference signal that will add to any desired signal. Our goal in this chapter is to find ways of keeping interference currents from flowing in any input signal conductor. To remove this coupling, the shield connection to circuit common must be made at the point, where the circuit common connects to the external ground. This connection is shown in Figure 4.3b. This connection keeps the circulation of interference currents on the outside of the shield.

There is only one point of zero signal potential external to the enclosure and that is where the signal common connects to an external hardware ground. The input shield should not be connected to any other ground point. The reason is simple. If there is an external electromagnetic field, there will be current flow in the shield and a resulting voltage gradient. A voltage gradient will couple interference capacitively to the signal conductors.

N.B.

An input circuit shield should connect to the circuit common, where the signal common makes its connection to the source of signal. Any other shield connection will introduce interference.

N.B.

Shielding is not an issue of finding a "really good ground." It is an issue of using the right ground.

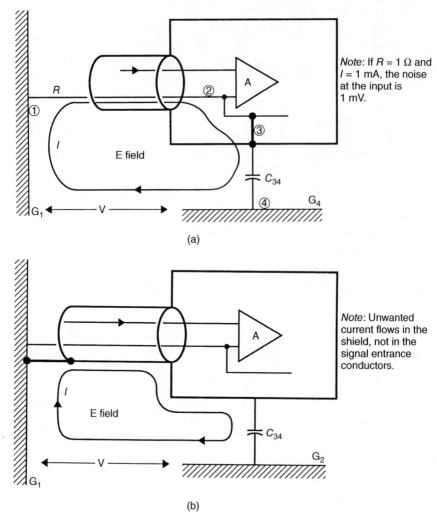

Figure 4.3 (a) The problem of bringing one lead out of a shielded region. Unwanted current circulates in the signal lead 2. (b) The E field circulates current in the shield, not in the signal conductor.

4.5 THE ENCLOSURE AND UTILITY POWER

When utility power is introduced into an enclosure, a new set of problems results. The power transformer couples fields from the external environment into the enclosure. The obvious coupling results from capacitance between the primary coil and the secondary coil. Note that the secondary coil is connected to the circuit common conductor. The

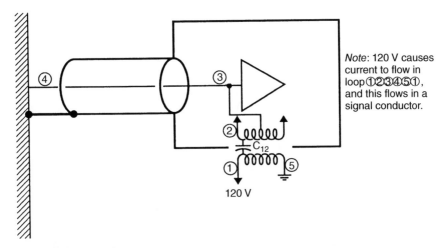

Figure 4.4   A power transformer added to the circuit enclosure

basic offender is the 120 V on the *ungrounded* or "hot" conductor. The reactive coupling is shown in Figure 4.4.

Unwanted current now flows in the loop involving the utility ground, the primary voltage, and the input signal common. If an ac adapter is used, the 60 Hz coupling is reduced. Without a filter, high-frequency interference can still couple into the input signal conductors. This interference pays little attention to turns ratios.

The construction of power transformers usually begins by winding a primary coil on a bobbin. After a layer of insulation, the secondary coil is wound over the primary coil. This practice places one end of the primary coil next to one end of the secondary coil. Typically, the interwinding capacitance is a few hundred picofarads. At 60 Hz, this is a reactance of about 10 MΩ. If the "hot" lead is next to the secondary coil, the resulting current at 60 Hz is 12 μA. For many applications, this level of current flow is not a problem. Of course, if the *grounded*[1] conductor were wound next to the secondary coil, the current flow at 60 Hz would be smaller.

The issue is often not the current flow at 60 Hz, but the noise current flow at higher frequencies. This noise can originate from three basic sources: The electromagnetic fields in the area, pulses, or signals on the power line generated by nearby hardware and *neutral* voltage drops. If the input common conductor is long, current flow in the conductor impedance can result in a significant voltage drop. This voltage drop is a normal-mode signal that adds to the signal of interest. In some cases,

---

[1] Terms that are defined by the NEC are shown in italics. See Section 5.3 for definitions.

power line voltage spikes on signal conductors can be great enough to damage unprotected hardware. Power line filters can limit this type of interference.

Out-of-band currents that flow in the input common conductor can generate signals that can interfere with the performance of both analog and digital hardware. Transient bursts can overload analog electronics resulting in offsets or "pops." If the out-of-band signal is a steady-state carrier, then there may be signal rectification resulting in dc offsets. To limit this coupling, small rf filters should be placed at the input to analog amplifiers. A typical low-pass filter might be a series 100-Ω resistor and a shunt 50-pF capacitor located at each input base or gate.

At this point in our development, there are two circuits entering the enclosure. These are the input common conductor and the power transformer. These conductors are associated with external grounds. The power transformer now circulates current to ground in the output common conductor. This current rarely causes problems because of the following:

1. Output signal levels are usually greater than a few volts.
2. The voltage drop in the output common is not amplified.
3. The signal output impedance is low.
4. Output cable runs are usually relatively short.

The output signal common can be inside a shielded cable, or it can be the shield itself. In this latter case, the shield should be treated like a signal conductor, not an extension of the enclosure. If a two-conductor shielded cable is used, the shield can be terminated (grounded) at one or both ends. The preferred connection is where the signal common terminates. It is acceptable to transport the output signal on open wires, but it is rarely done.

4.6 THE TWO-GROUND PROBLEM

The circuit in Figure 4.4 has one grounded input conductor plus a power transformer connection. If the input and output circuit common are both brought out of the enclosure and grounded, the result is the familiar "ground loop." The fields in the external environment will couple to the loop formed by the common conductor connecting between two ground points. Currents flowing in this loop can flow in the transducer source impedance. In many instances, the interference created by this ground loop can be larger than the signal of interest.

Ground loops can result when signal cables connect to circuit commons that are in turn connected to *equipment ground*. For example, in bench testing, the item under test and an oscilloscope form a ground loop. It is a good testing procedure to isolate the oscilloscope with a "cheater plug" to avoid this loop. This plug simply breaks the *equipment ground* connection provided by the manufacturer. This third-wire safety connection is required by the NEC. If an isolated oscilloscope is connected to the "hot" or *ungrounded* power line the shock hazard is obvious. It is good practice to place a warning label on equipment that is used un*grounded*.

There are many systems made up of interconnected pieces of hardware. These systems may be analog or digital in nature. If the hardware is rack mounted, then the *equipment grounds* connect the hardware together. For digital hardware, most of the problems discussed in this chapter will not be an issue because the signal levels are several volts, not millivolts.

4.7 INSTRUMENTATION AND THE TWO-GROUND PROBLEM

The basic analog problem is to condition a signal associated with one ground reference potential and transport this signal to a second ground reference potential without adding interference. Consider the arrangement in Figure 4.5. The input and output circuits have been separated so that the input common grounds at the source of signal and the output common grounds at the termination of the output signal. This unbalanced signal source represents the most difficult problem in instrumentation. The ground potential difference between the two enclosures causes current to flow in the unbalanced source resistor

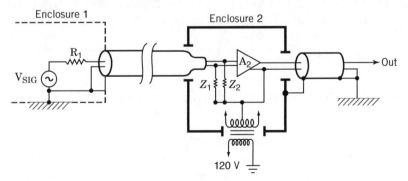

Figure 4.5 The two-circuit enclosures used to transport signals between grounds

R_1 and the impedance Z_1 in enclosure 2. The current is limited by the values of R_1 and Z_2. Note that Z_1 and Z_2 are actually the internal impedances of A_2. They are brought out as components in the figure to make a point.

The balanced signal source is usually a strain gauge. This type of signal is discussed in Section 4.8. Low-impedance sources such as thermocouples are discussed in Section 4.10.

Input signal conductors should be guarded right up to the input base or gate of the input amplifier. The guard shield in this arrangement is not a circuit conductor. It is an electrostatic shield. This shielding must be carried into the instrumentation on an individual channel basis. On a circuit board, a conducting surface or added traces can guard the signal instead of a cable shield.

In this approach, the high input impedance differential amplifier provides all the needed gain. There is circuitry between the input leads and the actual amplifier input. This detailed treatment of the input circuit is not shown in the figure. This circuitry is really a part of the input enclosure, so it must be well guarded. It includes diode clamps, high-impedance conductive paths for gate or base currents, and any input rf filters. This circuitry is needed but usually it is not a part of any specification. This circuitry is the responsibility of the manufacturer. The buyer should know that this protection is needed.

Consider instrumentation in Figure 4.5, where the gain is 1000 and the input-unbalance resistance R_1 is 1000 Ω. If the output error limit is 10 mV, the input error is 10 μV; the current in the unbalance resistance is limited to 10 nA. If the common-mode voltage is 10 V, the impedance Z limiting common-mode current flow must be 1000 MΩ.

General-purpose instrumentation is designed to have an input impedance of 1000 MΩ on both inputs. This way the unbalanced source resistance can be on either input. This requirement is what makes general instrumentation design so difficult.

N.B.

1000 MΩ is the reactance of 2 pF of capacitance at 60 Hz.

The ability to reject a common-mode signal is called the common-mode rejection ratio or CMRR. If a common-mode signal is 10 V and the resulting output signal is 10 mV the rejection ratio is 1000 to 1 or 60 dB. If the amplifier gain is 100, the error signal at the input is 0.1 mV. The ratio of 0.1 mV to 10 V output is 100,000 to 1 or 100 dB. This figure is the CMRR referred to the input (rti). In the previous example, the CMRR referred to the input is a million to 1 or 120 dB measured at 60 Hz with a 1000-Ω input line unbalance.[2]

To maintain a reactance of 1000 MΩ at 600 Hz, the leakage capacitance would have to be held to 0.2 pF. These numbers show how difficult it is to reject higher frequency common-mode signals using this type of circuitry. If the signal is somehow radiated or sent over an optical link, then this high input impedance is not needed. The problem is then shifted from providing a high input impedance to modulating a carrier with great signal accuracy. If the bandwidth must be 100 kHz and the signal errors must be kept under 0.1%, this represents a difficult problem.

N.B.

A guard shield should be connected to the input signal conductor at the point where the signal connects to the external reference point. In a multichannel system, the best practice is to provide each signal with its own guard shield.

The input guard shield is necessary in areas where there are nearby conductors that are not at the input ground potential. In Figure 4.5, the guarding near R_1 may not be needed as all nearby conductors are at the input reference potential. Inside enclosure 2, most of the conductors are referenced to the output common. In this area, the input guard shield must surround the input signal line. Any leakage capacitance is a part of the input impedances Z_1 or Z_2.

The technique we have just discussed uses a differential amplifier that is located a distance from the transducer. When amplification is provided at the transducer, then several different approaches are possible:

a. Provide an rf link using a modulator and demodulator (analog or digital).
b. Provide optical coupling using a digital data link.
c. Use a current loop to couple signal to a remote differential amplifier.

The first two methods raise the isolating impedance Z to infinity. These methods can be used to transport signals from an aircraft to earth or between two buildings or between computers. Current-transmitting

---

[2] See Appendix A for a discussion on decibels.

loops are used in many industrial applications, where a wide bandwidth is not needed. The differential method shown in Figure 4.5 is often used in general-purpose test bays. When all of the electronics is located in the test bay, it is convenient to reconfigure the testing by moving transducers. Providing electronics at each transducer or at groups of transducers modifies the structure and is often not practical.

Amplifying the signal near each transducer can help solve the common-mode interference problem. If the full-scale signal level is increased to 100 mV or greater and the new signal source impedance is below a few ohms, the remaining common-mode interference problem is not as difficult to solve. If the pre-gain is 100, then the CMRR (common-mode rejection ratio) in the second amplifier may only have to be 10^3. This approach is practical when there is no need for signal conditioning at the transducer. Any reduction in the required CMRR does not change the requirement for a high Z in the common-mode path. This high impedance is needed to limit the common-mode current flow in any input imbalance just as before.

4.8 STRAIN-GAUGE INSTRUMENTATION

Strain-gauge instrumentation is a two-enclosure problem as shown in Figure 4.6. The signal source is a symmetric Wheatstone bridge. The input enclosure contains the gauge resistors as well as the source of transducer excitation. In this diagram, all four bridge arms are active, which means they are mounted on the structure being tested. In many applications, only one arm of the bridge is active and the remaining arms

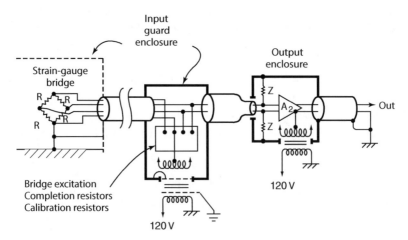

Figure 4.6 The basic strain-gauge circuit

are located near the source of excitation. All bridge arms must be inside the input guard enclosure. If the excitation source is centertapped, then only two bridge resistors are required.

The capacitance from a conducting surface to a gauge element can be several hundred picofarads. It is safe to assume that there will be a potential difference between the surface under test and the instrumentation output ground. If the gauge is not grounded to the surface under test, this potential difference will add signal to the input circuit. The interference will be the result of current flow in the parasitic capacitance between the gauge arm and the structure. If there is only one active arm, the interference will be the greatest. The best protection against this type of interference is to connect one corner of the bridge to the structure and connect the guard shield to this same point.

A single strain-gauge bridge can require up to 10 conductors. These conductors provide for excitation, remote excitation-level sensing, calibration, a signal pair, and the guard shield. The guard shield should be carried through any interface without a break. Connecting all the guard shields together at an intermediate point can introduce interference. Connecting segments of the guard shield to different grounds is also incorrect. If this practice is required, some testing is recommended to determine the level of interference. Compromise is a way of life. It is important to understand when a compromise has been made.

When the gauge elements of the bridge undergo stress or strain, the unbalanced resistance can be as high as 50 Ω. If the interference error limit is $10\,\mu V$, the interference current flow is limited to $0.2\,\mu A$. If the common-mode voltage is 10 V, the impedance Z allowing common-mode current flow must be $5\,M\Omega$. This is the reactance of 3 pF at 10 kHz. This mode of interference coupling is maximum for a full-scale signal. In effect the stress or strain modulates the interference. This problem can only be resolved by limiting common-mode current flow in the gauge arms.

N.B.

A resistance strain gauge generates millivolts of signal. Errors that are proportional to signal level are not acceptable.

4.9 THE FLOATING STRAIN GAUGE

An electrical connection from a strain gauge to the structure is not always convenient. If the structure is not used as the reference conductor, then there can be current flow from the structure through leakage capacitance through the gauge elements to the input cable shield. If the coupling is symmetrical, then this current flow creates a common-mode signal that is rejected by the amplifier. If one arm of the gauge is active, then the reactive coupling is definitely unbalanced and any coupled interference will be normal mode. This signal is amplified – not rejected.

The input guard shield and one side of the signal are shown connected to the structure in Figure 4.6. If there is no connection to the structure, the guard shield should still be connected to the low side of the excitation. In a relatively quite environment and because the strain gauge is basically a balanced system, the structural connection of the shield and the signal as shown in Figure 4.6 may not be required.

Most instrumentation provides an internal path for input base or gate currents. This allows the instrument inputs to be left open without causing the instrument to overload. These paths are often in the 100-MΩ range. Even if these paths are provided, it is still best to avoid making measurements, where the input leads have a high impedance to ground. It is possible for the inputs to saturate when there is no path for current to flow. When this happens, the inputs are in overload but because of feedback the problem may not be noticed. It is worth noting that floating the input is usually not a part of an instrument specification.

Making measurements on a floating structure is also not recommended. If possible, a grounding strap should be used so that the potential difference is controlled. One reason is that it is very difficult to mount a gauge and control the coupling capacitance to the structure. If one gauge element is mounted closer to the structure than another, the reactive current flowing in the gauge resistances will not be symmetrical. In a noisy environment, this lack of symmetry will introduce a normal-mode signal. This interference can only be rejected by limiting the bandwidth of the system.

In applications where it is difficult to ground the gauge and guard shield to the structure under test, one central point can be used as a ground for a cluster of gauges. This works in an electrically quiet environment.

N.B.

A high input impedance is needed to reject common-mode signals. This impedance does not imply that operating from high-impedance signal sources is acceptable.

There are actually two common-mode signals that must be rejected by an instrument amplifier. We have just discussed the ground potential difference mode. The second common-mode signal is one half of the excitation supply. If the excitation level is 10 V, then this common-mode signal is 5.0 V. The CMRR for this signal should also be 120 dB. As an example, if the excitation level is 10 V, the error signal or offset that is introduced should be less than $5\mu V$. This requirement is usually not mentioned in a list of specifications. Since the excitation level is constant, this is a static specification.

4.10 THE THERMOCOUPLE

A thermocouple is formed by joining two dissimilar metals and bringing them back as conductors to a reference temperature. The voltage between the two conductors at the reference temperature is a measure of the temperature at the junction. This reference temperature can be an ice bath or a temperature-controlled surface. If a reference temperature is not used, a temperature compensation circuit can be used in the instrumentation. This compensation circuit corrects the voltage based on the ambient temperature. In either case, the resulting voltage is a measure of the junction temperature. At the reference point, the dissimilar metals are connected to copper wires that carry the measured voltage to the instrumentation.

The thermocouple junction is often bonded to a conducting surface to obtain a good measure of temperature. Theoretically, this bonding point is where the input guard shield should tie. In practice, the guard shield is usually connected where the thermocouple conductors transition to copper. This is satisfactory because the input unbalance resistance is low and bandwidth is usually not of concern.

If the thermocouple is used to measure the temperature of a fluid, then the junction does not contact a conducting surface. The input signal guard shield should still connect to one side of the signal. One solution is to connect the shield to one of the thermocouple leads at the transition point to copper. The input leads should not be left floating because there is a chance of overload.

Because of possible aliasing errors, it is good practice to filter signals before any digital sampling. For a floating thermocouple, the midpoint of a balanced RC input filter can be used as a grounding point. This type of filter attenuates both normal and differential mode noise. The impedance of the RC filtering circuit can be $1.0\,k\Omega$. If the capacitors are $0.1\,\mu F$, the cutoff frequency will be 1.59 kHz. A higher resistance can be effective, but the dc drift of the instrumentation might limit accuracy. Aliasing errors are discussed in Section 4.22.

N.B.

If the data is the same when analyzed at different sample rates, then aliasing errors are not involved.

4.11 THE BASIC LOW-GAIN DIFFERENTIAL AMPLIFIER (FORWARD REFERENCING AMPLIFER)

A simple low-gain differential amplifier is shown in Figure 4.6. This type of amplifier can be applied when the input signal has a low source impedance and signal levels are above 0.1 V full scale. The circuit has inputs labeled V_{IN1} and V_{IN2}. The gain from V_{IN1} to the output is $+R_2/R_1$. The gain from V_{IN2} is $-R_2/R_1$.

If the same signal is applied to the two inputs, the gain is near zero. If a difference signal V_{DIFF} is applied between the two inputs, the output signal is $V_{DIFF} R_2/R_1$. This circuit provides gain to a difference signal and rejects the average or common-mode signal. If either input is at 0 V (grounded) the gain to the other input is the ratio of resistors R_2/R_1. The sign of the gain is plus or minus depending on which input is used.

The differential amplifier can be used as the input circuit for the second enclosure as shown in Figure 4.7. The added differential amplifier uses the power supplies available in the second enclosure.

If the output common of the second enclosure is taken as the reference conductor, then the input ground potential of the first

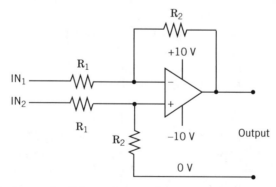

Figure 4.7 The low-gain differential amplifier applied to the two-ground problem

enclosure becomes the common-mode signal. Ideally, the gain for the common-mode component of the signal should be zero. In this application, the differential amplifier is called a forward referencing amplifier. This amplifier re-references the signal found in the first enclosure to the signal common found in the second enclosure.

The common-mode rejection of the circuit in Figure 4.8 depends mainly on the ratio of feedback resistors. If the resistors are equal and their ratios are matched to 1%, the CMRR will be about 100:1. This means a 1-V common-mode signal at 60 Hz will generate a 10 mV error. This is a 0.1% error compared to a 10 V output. Typical resistor values might be $10\,k\Omega$. This same circuit can be applied to isolating analog video signals if the integrated circuit amplifier has adequate bandwidth. In this application, the feedback resistors should be about $1\,k\Omega$. In analog video applications, the signal is usually limited to 2 V peak-to-peak.

N.B.

Source impedances must be considered when common-mode rejection depends on having an accurate ratio of feedback resistors.

The gain of the differential amplifier in the above example is unity. If gain is provided by the differential amplifier or by circuits that follow, a higher CMRR should be provided. For example, if the gain in the amplifier is 10, the CMRR should be 1000:1 to limit the signal error to 0.1% of full scale.

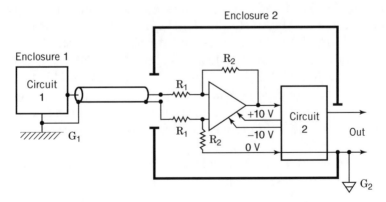

Figure 4.8 The basic low-gain differential amplifier

A CMRR depends on the feedback factor (loop gain) of the integrated circuit. See Section 4.18. At frequencies above a few kilohertz, the CMRR will generally fall off proportional to gain. It is a good design practice to verify the CMRR at high frequencies if this performance is needed.

The resistances R_1 and R_2 in Figure 4.8 limit the amount of common-mode current that flows in the input common conductor. If this value is $10\,k\Omega$ and the common-mode voltage is 1 V, the current in R_1 is 0.2 mA. If the input common lead has a resistance of $1\,\Omega$, the interference coupling is 0.2 mV. This coupling has no relationship to the CMRR of the forward referencing amplifier. If the feedback resistors are $100\,k\Omega$, the current flow would be reduced by another factor of 10.

Common-mode rejection does not reduce coupling from the power transformer. To limit this coupling, shielding can be added to the power transformer. This shielding is discussed in the following section.

4.12 SHIELDING IN POWER TRANSFORMERS

A basic transformer shield consists of a single wrap of foil between coils. This foil must be insulated at the overlap to avoid a shorted turn. An effective shield can be made from a thin layer of either copper or aluminum. A connection to the shield can be made by taping a bare wire to the shield or by soldering to an installed copper eyelet. The shield lead is usually brought out for an external connection. A shield in a 10-W transformer might limit the mutual capacitance from the primary coil to the secondary coil to about 5 pF. This shield is shown in Figure 4.9.

A single shield between the primary and secondary coils can help to limit reactive current flow in the input signal common. The proper connection for this shield is to *equipment ground*. If the shield were connected to the enclosure or to the signal common, the power fault path would be through the input cable and this is not acceptable. A single shield cannot limit the current flowing in the loop ① ② ③ ④ ⑤ ⑥ ① or in the loop ⑦ ③ ④ ⑤ ⑥ ⑦. It would take two additional shields to control this current. The single shield helps when there is a potential difference between the *equipment ground* and the *grounded* or *neutral* power conductor. This potential difference is often the result of harmonic current flow in the facility *neutral* conductor.

A power transformer used inside of an instrument would ideally require three shields to limit most of the unwanted power current flow. The shield next to the primary coil would connect to *equipment ground*. The center shield would connect to the enclosure and the shield around

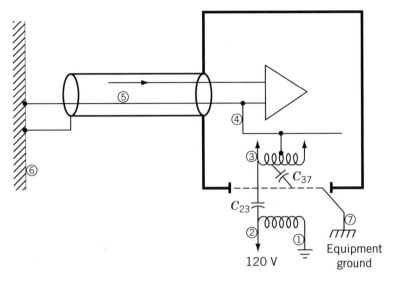

Figure 4.9   The single shield applied to a power transformer

the secondary coil would connect to circuit common. An example of two shields is shown in enclosure 1 of Figure 4.6. To limit the mutual capacitances to around 0.2 pF, the shields would have to "box" the coils. (A "box" shield fully encloses each coil.) These shielded transformers must be handcrafted and as a result they are expensive. "Boxed" shielded transformers are rarely used in today's electronics.

There are applications where the circuit common grounds through a series resistance. An application might be a floating power supply. Three shields as mentioned above can be used to control the flow of reactive currents in the transformer mutual capacitances. In a transformer with three "boxed" shields, the leakage capacitances can be held to approximately 0.2 pF. At 60 Hz, this is a reactance of $13 \times 10^9 \, \Omega$. The circulating current at 120 V, 60 Hz is about 9.2×10^{-9} A. This is a voltage drop of about 9.2 μV in a resistance of 1000 Ω. Fortunately, there are circuit techniques that can float a power supply without the use of these expensive multishielded transformers.

4.13 CALIBRATION AND INTERFERENCE

There are many approaches to calibrating instrumentation. Bench calibration can test for parameters that online testing cannot measure. Online calibration just prior to a test can verify basic operation and provide data to correct for a few of the errors.

N.B.

No amount of calibration can eliminate errors caused by interference.

Measurement errors include factors such as linearity, gain, offset, amplifier input noise, common-mode rejection, temperature coefficients, excitation accuracy, signal losses in a cable, rise time, and settling time. Calibration can correct for errors in gain and offset but not for errors caused by common-mode signals. Specifications often call for total error levels of 0.1% of full scale. This is very difficult to verify.

N.B.

A microampere of unwanted current flowing in 1 Ω generates 1 μV of error.

4.14 THE GUARD SHIELD ABOVE 100 kHz

The guard shield should protect the input signal up to the input bases or gates. The presence of the guard shield in the instrumentation can couple high-frequency fields into the enclosure. Even if these signals are out of band, they can cause errors that result from overload and signal rectification. It is good practice to connect the shield to the enclosure at frequencies above 100 kHz through a series RC circuit. The connection should ideally be located outside of the instrument, but it is usually placed at or near the connector. Typical values are $R = 100$ Ω and $C = 0.01\,\mu\text{F}$. This circuit limits the high-frequency fields that enter the enclosure. This RC circuit is shown in Figure 4.10.

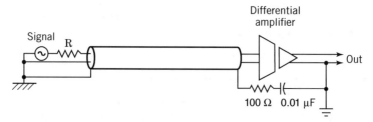

Figure 4.10   The RC bypass on the input guard shield

4.15 SIGNAL FLOW PATHS IN ANALOG CIRCUITS

The basic components that make up an analog circuit are integrated circuits, feedback resistors, resistive voltage dividers, diode clamps, zener diodes for voltage limiting and RC filter components, and a few power supply filter capacitors. Typically, these components are interconnected on a two-sided printed circuit board. The dc power is supplied from a nearby rectifier system and/or voltage regulators. On some circuit boards the power is supplied locally from a power transformer. Note there is usually no ground plane available when using a two-layer board. If care is taken in layout, a ground plane is not needed. The electronics may still require an enclosure connected to the input guard shield. External signal and power connections can be made using connectors or solder pins. When digital circuits are involved, a ground plane may be a necessity.

Here are a few rules that will help in analog board layout.

1. Maintain a flow of signal and signal common from input to output. The area between the signal path and the signal reference conductor should be kept small.
2. Components associated with the input should not be near output circuit components.
3. Power supply connections (dc voltages) should enter at the output and thread back toward the input. This avoids common-impedance coupling (parasitic feedback).
4. The greatest attention should be paid to the input circuit geometry. Lead length for components connecting to the input path should be kept short. Another way of describing this requirement is to interconnect the components to minimize the amount of bare copper connected to the input signal path.
5. Feedback summing points are critical. Keep lead lengths short at these nodes.

4.16 PARALLEL ACTIVE COMPONENTS

When transistors or field effect transistors (FETs) are paralleled for added performance, there is a good chance for an instability. If an oscillation should take place, it can be high enough in frequency to go undetected. An oscillator of this type can overheat components and/or limit

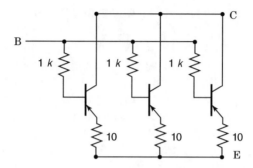

Figure 4.11   Adding suppression resistors to parallel circuit elements

their effective gain. Radiation from this type of oscillator can interfere with nearby circuits. If the circuit is marginally stable, the oscillations might occur after a long warm-up or for some values of load. It is good practice to place a resistor in series with each emitter and add a series resistor to each base. Typical emitter resistors might be 10 Ω and typical base resistors might be 1000 Ω. These resistors are often called suppression resistors. A typical circuit is shown in Figure 4.11.

N.B.

If more than one element of a component are paralleled to a second component, a series resistor should be used.

4.17 FEEDBACK STABILITY – INTRODUCTION

Integrated circuit amplifiers are usually supplied with a very high forward gain at dc. Negative feedback is usually required to make these devices practical. This type of amplifier is supplied with internal compensation, which means there is internal shaping of the open-loop frequency response. Without this compensation, feedback will usually result in oscillation. Even with compensation, there are applications where circuit stability can be marginal. To understand the problem, a brief review of feedback theory is needed. Because stability is not guaranteed, it is a good idea to test every feedback circuit to make sure it is unconditionally stable.

Here are some definitions that will help in understanding this section:

1. *Negative feedback*. Subtracting a fraction of an output signal from an input signal and amplifying the difference.

2. *Open-loop gain.* The gain before feedback.
3. *Closed-loop gain.* The gain after feedback is applied.
4. *Feedback factor.* The forward gain in excess of the closed-loop gain.
5. *Positive feedback.* Adding a fraction of the output signal to reinforce the input signal. This usually results in an oscillator.

4.18 FEEDBACK THEORY

A basic feedback circuit is shown in Figure 4.12. The input signal is the sum of a fraction of the output signal βE_{OUT} and the input signal E_{IN}. This summation usually takes place in a differential input stage or by a resistive divider. The gain of this feedback circuit is

$$\frac{E_{\text{OUT}}}{E_{\text{IN}}} = -\frac{A}{(1 + A\beta)}. \tag{4.1}$$

If A is negative and large, then the gain is very close to $-1/\beta$, which is the ratio of resistors in Figure 4.13.

In an internally compensated amplifier, the forward gain falls off proportional to frequency. The bandwidth of the closed-loop amplifier is approximately the frequency where the open-loop gain is equal to $1/\beta$. As an example, consider an amplifier with an open-loop dc gain of -10^6. If the closed-loop gain is -100 and the bandwidth is 100 kHz, the open-loop gain of the amplifier had to start down (losing bandwidth) around 10 Hz.

The phase shift associated with any circuit is closely associated to the attenuation slope. For a slope that is proportional to frequency, the

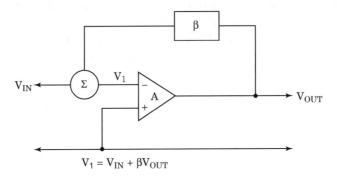

$$V_1 = V_{\text{IN}} + \beta V_{\text{OUT}}$$

Figure 4.12   The basic feedback circuit

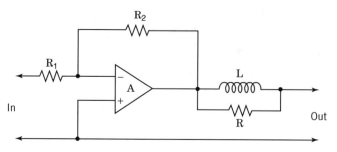

Figure 4.13 An LR-stabilizing network

phase shift is 90°. For a slope that is proportional to the square of frequency, the phase shift is 180°. In the compensated feedback amplifier
of Figure 4.11, the phase shift of the open-loop amplifier is approximately 90° over the range 10 Hz to 100 kHz. The phase shift of the
closed-loop amplifier is 90° divided by the feedback factor. For example,
if the closed-loop gain is 100, the feedback factor at 1 kHz is 10^2. This
means the phase shift at 1 kHz is approximately 0.9°. At 100 kHz, the
feedback factor is unity and the phase shift has increased to about 45°.
At frequencies above 100 kHz where the attenuation slope is proportional to frequency, the phase shift will be 90°.

In Equation 4.1, the gain A must not have a phase shift greater than
180° before $A\beta$ reaches unity. If this condition is not met, the circuit will
oscillate. This condition is known as the Nyquist criteria. If the phase
shift approaches 180° as $A\beta$ approaches unity, the result is an amplifier
with a very large peak in its amplitude/frequency response. The transient response of this amplifier will have a large overshoot with many
cycles of ring-down. This is an indication that the circuit is marginally
stable. This ringing condition indicates there is a problem that needs
correction.

Feedback systems have limited gain to any internal interference. If
a 60-Hz signal is injected at an internal point, the gain to this signal is
the closed-loop gain reduced by the gain ahead of the injection point. If
the closed-loop gain is 100 and the gain preceding the injection point is
1000, a 0.1 V interfering signal would by multiplied by 100 and divided
by 1000. The result would be an output signal of 0.01 V. If the output
stage has a linearity error of 1%, this error is reduced by the feedback
factor. If the feedback factor is 100, the resulting linearity error would
only be 0.01%.

4.19 OUTPUT LOADS AND CIRCUIT STABILITY

If the feedback amplifier in Figure 4.12 is connected to a capacitive load, the result will be additional phase shift in the forward open-loop gain. A capacitive load might be a signal cable having a capacitance of several hundred picofarads. When feedback is applied around this block of gain, an instability can result. The problem is most severe if the closed-loop gain is unity. It is good practice to place a parallel LR circuit in series with the output if this output is intended for general use. The L can be 10 μH and the resistor 10 or 20 Ω. This circuit is shown in Figure 4.13. At frequencies above 100 kHz, the output impedance is the resistor in series with the output impedance.[3] This resistance is usually sufficient to avoid any instability. If a 10- or 20-Ω resistance in series with the output poses no problem, then the inductor is not necessary.

It is good practice to test every output circuit for stability. The preferred test is as follows: A square wave should be used to drive the output through a 100-Ω series resistor. If there is ringing at the output terminals, the circuit should be revised. Note the drive signal must not overload the output circuit. For this reason, it is good practice to observe stability using a small amplitude square wave signal. Large signals can sometimes reduce the loop gain and the ringing (instability) may not be apparent. A resistive load applied to the output of a feedback amplifier can reduce the open-loop gain. This can often provide a margin of stability. For this reason, stability tests should be made with the output unloaded.

4.20 FEEDBACK AROUND A POWER STAGE

If an integrated circuit amplifier drives a power output stage, it is desirable to include the power stage in the feedback loop. The added phase shift (delay) in this output stage can cause instability. A simple way to

---

[3] The output impedance of an amplifier with negative feedback is the impedance of the output stage reduced by the feedback factor. Because the feedback factor falls off linearly with frequency, the output impedance rises with frequency. A rising output impedance means the output impedance looks inductive. Any capacitive loading is in series with this inductance represents a series resonant circuit. If the phase shift at the feedback sensing point ever reaches 180° and the gain is greater than unity, the circuit will oscillate.

avoid this problem is to provide a feedback path at high frequencies directly to the output of the integrated circuit. This arrangement is shown in Figure 4.14. At high frequencies, the feedback path is through C_1 and R_1.

4.21 CONSTANT CURRENT LOOPS

Instrumentation is available that transmits signal data over long lines using a constant current source. The standard range of current is 4–20 mA. The 4 mA current value can represent the zero of signal. A constant current source implies that the current flow is independent of the load and loop resistance or any external voltage induced in the current loop. A precision series resistor can be used to convert the current to a voltage at the receiver. This voltage is then amplified by a differential amplifier located in the second enclosure. If the loop is opened for any reason, there is no feedback. Under this condition, the circuit will overload.

N.B.

A current loop implies a high source impedance.

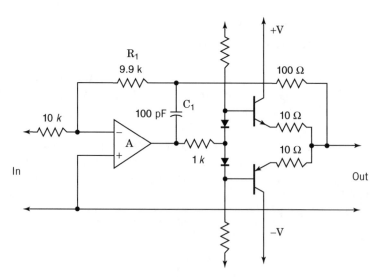

Figure 4.14 Feedback around a power stage

4.22 FILTERS AND ALIASING ERRORS

Signals that are sampled must be filtered to avoid what are called aliasing errors. If the sample rate is 10 kHz, there can be no signal content in the data above 3 kHz. As an example, an 8 kHz signal sampled at 10 kHz will result in a 2 kHz output. Note that 2 kHz is the difference between 8 kHz and the sample rate of 10 kHz. This problem does not exist if the sample rate is high enough.

N.B.

In movie Westerns, stagecoach wheels may appear to rotate backward. This is a form of aliasing error. To eliminate this effect, the image scan rate would have to increase.

Recording test data digitally is inexpensive. To limit aliasing errors, at least three samples must be taken of the highest frequency of interest. It is interesting to consider the issue of thermocouples. Temperature does not usually change rapidly except possibly in an explosion or for a missile impact. The question is whether the sampling rate can be 30 Hz when the bandwidth of the amplifier is 100 kHz. Technically the answer is no. Analog filters could be used to reduce the amplifier bandwidth to 10 Hz, so the sample rate can be 30 Hz. This would be a proper solution. For thermocouples, it makes more sense to avoid filters, use the 10 kHz bandwidth, and sample the data at 30 Hz. The data can be analyzed using different sample rates at say 2, 5, and 15 samples per second. Character differences in the result would be an indication of an aliasing error.

In general instrumentation, the full nature of the signal is often unknown. If the sample rate is set, the bandwidth of the amplifier being sampled must be limited. This band limiting must be done by using analog filters. Unfortunately, this filtering cannot be done digitally. The active filters that are often used have a very sharp cutoff character. The attenuation rate above the cutoff frequency might be 60 dB per decade. An antialiasing filter that has a 60 dB-per-decade slope must start attenuating the signal at 500 Hz to provide 60 dB of attenuation at 5 kHz. These analog filters can take on many forms. Elliptical filters are often used to provide a flat frequency response before there is attenuation. This type of filter has a significant transient overshoot. Bessel-type filters have a very soft knee with no transient overshoot. The type of filter selected should match the type of data that is expected. Noisy data can excite transient overshoot in a steep filter and this can result in signal overload. Unfortunately, these analog filters add cost to the instrumentation particularly if they are adjustable.

N.B.

Digital filtering is practical once the data has been correctly sampled.

N.B.

If the data sample rate is reduced, the analog aliasing filter cutoff frequency must also be reduced by the same factor.

4.23 ISOLATION AND DC-TO-DC CONVERTERS

The transformers used in dc-to-dc converters provide a degree of isolation not provided by 60 Hz transformers. These converters are not free from problems however. Converters are smaller and less expensive than 60 Hz transformers and for this reason they have found wide acceptance in electronics hardware design.

In a typical dc-to-dc converter power supply, utility power is rectified and energy is stored in an electrolytic capacitor. This stored energy is modulated at a frequency above 50 kHz and a transformer couples stored energy to the circuits of interest. The modulator produces a square wave voltage that is rectified in the secondary environment. Energy is supplied to the transformer once per modulation cycle and this means that large energy storage capacitors are not needed on the secondary. The problems with these converters have to do with coupling across the transformer at the fundamental and harmonics of the modulation frequency. Specifically, the leading edges couple voltage spikes across the transformer.

Shielding in a modulation transformer is not practical. The capacitances that would be introduced would make the circuit unusable. If the primary peak voltage is 170 V, the square waves on the primary coil can be 340 V peak-to-peak. If the rise time is 1 µs, the current flowing between the primary and secondary coils depends on the mutual capacitance between the coils. If this capacitance is 10 pF, the current spike has a peak amplitude of 3.4 mA. This magnitude of current flowing in an analog circuit common is generally not acceptable.

There are several ways that this current spike can be reduced. If the transformer is built with a centertapped primary (so that there are positive and negative going voltages), then equal and opposite voltages are coupled through equal capacitances to the secondary coil. Often bifilar wound coils can provide this balance. Under these conditions, the currents can cancel. This cancelation is not perfect, but the currents can be reduced by a factor of 10–100. Sometimes, a trimming capacitor might be necessary. If the square wave of voltage is generated with respect to *equipment ground*, any filtering of the current pulse must be with respect to this *ground*.

A way to reduce high-frequency coupling to a secondary circuit is to use two converter transformers in cascade. The second transformer is connected to a coil on the first transformer. This added transformer can be balanced (centertapped) and can operate at a low voltage. The coupling between the two transformers can be referenced to an output circuit common. The low voltage limits the high-frequency currents that might flow in an input common. This arrangement is shown in Figure 4.15.

The power line voltage can be rectified and energy stored in a capacitor. This energy can then be modulated and used for a power supply. During part of the cycle, energy for the modulator can come from the power line as well as from the storage capacitor. To limit this demand, a line filter is needed. This filter places capacitors between the power line and equipment ground. This means that some of the modulation current can enter the *equipment grounding* grid. The loop areas involved in this current flow are not well defined. The result is there can be fields generated by the modulator that creates interference for an entire facility. For this reason, a dc-to-dc converter should be designed so that this current is limited. In facilities with many pieces of electronic hardware, this noise current level can cause troublesome interference fields.

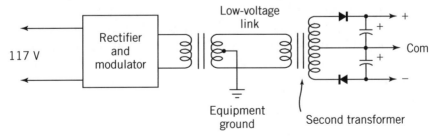

Figure 4.15   Using a second transformer to isolate switching noise

A step-down transformer is often used to reduce the ac voltage before rectification. In this situation, the dc-to-dc converter can operate at a lower voltage and the spike levels are reduced. The penalty is the cost and size of the step-down power transformer.

If care is taken, a dc-to-dc converter is a good way to generate isolated power supplies for transducer excitation. Multiple secondaries on a common transformer can be used to generate a group of isolated excitation supplies.

4.24 CHARGE CONVERTER BASICS

In vibration analysis, the sensing transducer is often a quartz crystal. This crystal is electrically equivalent to a capacitor. When the transducer is accelerated, a force is exerted across opposite faces of the crystal. This force generates a voltage on the faces of the crystal. This is called a piezoelectric effect. The acceleration can be measured by sensing the voltage or measuring the charge generated in the capacitance. The relationship between charge and voltage is $V = Q/C$, where C is the transducer capacitance. The relationship between charge and acceleration is a specification provided by the manufacturer.

The voltage on the transducer can be amplified by a high-impedance amplifier. The input cable capacitance attenuates the input signal and this makes calibration a function of cable length. The preferred method of amplifying signals from piezoelectric transducers is to measure charge generation not the voltage generation. The input cable capacitance does not attenuate the charge, making calibration much simpler. The charge is first converted to a voltage and the voltage is then amplified. This type of instrument is called a charge amplifier.

The basic feedback around an operational amplifier usually involves two resistors. The voltage gain is simply the ratio of the two resistors. If the resistors are replaced by capacitors, the gain is the ratio of reactances. This feedback circuit is called a charge converter. The charge on the input capacitor is transferred to the feedback capacitor. If the feedback capacitor is smaller than the transducer capacitance by a factor of 100, then the voltage across the feedback capacitor will be 100 times greater than the open-circuit transducer voltage. This feedback arrangement is shown in Figure 4.16. The open-circuit input signal voltage is Q/C_T. The output voltage is Q/C_{FB}. The voltage gain is C_T/C_{FB}. Note that there is essentially no voltage at the summing node.

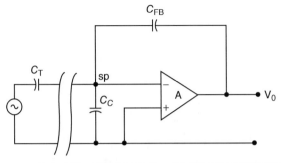

C_C = cable capacitance; C_T = transducer capacitance
C_{FB} = feedback capacitance
The summing point sp is a virtual ground.
The circuit transfers the charge generated
on C_T and places it on C_{FB}.

Figure 4.16   A basic charge amplifier

N.B.

A charge converter does not amplify charge. It converts a charge signal to a voltage.

The input cable adds capacitance between the summing point and signal common. The result is the noise generated by the input stage is amplified by the ratio of cable capacitance to feedback capacitance. Added input cable capacitance does not however change the ratio of input charge to output voltage.

If the transducer capacitance is $0.01\,\mu\text{F}$ and the feedback capacitor is $100\,\text{pF}$ the gain to an input voltage is 100. To provide a low-frequency response to 1 Hz the resistance across the feedback capacitor would have to be $10^{10}\,\Omega$. This can be provided by a 100-MΩ resistor and a feedback voltage divider of 100:1. This divider arrangement is shown in Figure 4.17. The path for input bias current is 100 MΩ. These high impedances require that the input stage to the amplifier must be a FET.

The noise level referred to the input for a charge converter is usually below 10^{-13} coulombs rms in 100 kHz bandwidth. Consider the coupling between the summing point and a nearby power supply voltage. If the coupling capacitance is 3 pF and the power supply has 30 mV of ripple, the noise coupled to the input is equal to 10^{-13} coulombs.

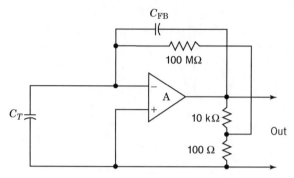

Figure 4.17 The resistor feedback arrangement to control the low-frequency response

To limit this coupling, the power supplied to a charge converter must be well regulated.

The smallest practical feedback capacitor in a charge converter circuit in Figure 4.15 is about 100 pF. One picofarad of capacitance represents a 1% gain error. Consider a gain of 100 following the converter. Capacitance from the summing point to this amplified signal can cause errors. If the gain is 100, the mutual capacitance must be held below 0.01 pF. This small value requires that the charge converter be carefully shielded electrostatically. Two small metal boxes that cover the components and circuit traces on both sides of a circuit board are usually adequate. The connection from the transducer to the charge converter must use special low-noise cable. This includes the signal connection inside any instrumentation. See Section 7.2.

N.B.

If the charge amplifier is a plug-in module, a dedicated coaxial input connector must be used. The input signal must be carefully shielded along its entire path.

N.B.

Transducers add weight to the item they are testing. A smaller transducer generates less charge.

The resonant properties of the transducer must be higher than any test frequency. If the transducer requires an insulating mount, this adds mass and another degree of freedom to the measurement. For high-frequency vibration testing, it is desirable to avoid this insulation. This means mounting the transducer common to the structure. This grounding requires that a differential amplifier stage be placed somewhere in the signal path. The stage can be a forward referencing amplifier in the terminating electronics, or it can be provided in the measuring instrument. In this case, the instrument must have two power supplies, one for the input charge converter and the second supply for the differential stage and any additional circuitry. The instrument would then be called a differential charge amplifier. Note the input signal is not balanced or differential in character.

4.25 DC POWER SUPPLIES

The design of dc power supplies has been left to last. The simplest dc power supply is a rectified voltage taken from the power line. A transformer is usually used to limit the voltage and to isolate the supply from utility conductors. The rectified energy is stored in an electrolytic capacitor. If the transformer is centertapped, then a plus and minus supply can be built.

Regulated power supplies make use of a standard integrated circuit. A range of voltages and currents are available. These supplies are in effect dc amplifiers with a fixed input signal derived from a zener diode. The feedback built into the amplifier produces a voltage source that has a low output impedance. It is customary to place an electrolytic capacitor at the output terminals so that there is no sensitivity to the character of the load. The integrated circuits provide internal protection against overload and high temperature.

4.26 GUARD RINGS

There are applications where the resistivity of a circuit board allows unwanted current to flow. A grounded ring can be placed around a key circuit point to divert this board current. A second solution involves a group of grounded vias located in a circle. If this protection is not adequate the circuit may have to be built on sapphire insulators.

4.27 THERMOCOUPLE EFFECTS

In dc amplifiers, the input conductors require special attention. Every joint or connection is a possible thermocouple. When every microvolt counts, the following rules apply:

1. Maintain circuit symmetry for both inputs.
2. Keep the input leads very close together.
3. Do not place any heat near the input leads.
4. Do not blow ventilation air across these leads.

4.28 SOME THOUGHTS ON INSTRUMENTATION

The basic problem in instrumentation is accepting the fact that there are so many reference commons. Typically a signal originates in one structure, is conditioned, and then terminated in a second structure. Each data channel must be separately powered to avoid instabilities and limit cross coupling. The standard approach provides one input power supply, an output power supply, and in some cases an excitation power supply per instrument. For a 10-channel system there are 30 reference commons. This fact is not going to change any time soon.

The digital world is big and it obviously overshadows the analog world. Once data is digitized, the need for reference conductors, dc drift correction, controlling cross coupling and treating aliasing errors all fade away. As this book shows, there are a host of new problems that must be considered in the digital world. If we had digital strain gauges and thermocouples, the measurement world would be so much simpler.

In the design of human beings, nature has found a way to be both analog and digital at the same time. The body functions in a continuous analog sense; yet messages are sent as pulses on neural networks. We have a core memory and despite the number of examples we can observe, the way it works is still poorly understood. It is self-correcting yet there are storage areas that are temporary. It is very efficient power wise.

CHAPTER 5

Utility Power and Facility Grounding

OVERVIEW

This chapter discusses the relationship between utility power and the performance of electrical circuits. Utility installations in facilities are controlled by the NEC (National Electrical Code). Safety and lightning protection requires that facilities connect their systems to earth. Designers of electronic hardware use utility power and also make electrical connections to earthed conductors. This sharing of the earth connection creates many problems that are considered in this chapter.

Ground planes and isolation transformers can be used to limit interference. The role of line filters, *equipment grounds*, and ground planes in facilities is explained. The problems associated with using *isolated ground* conductors are discussed. Lightning protection in facilities and for watercraft is a big safety issue. The fact that current cannot enter the water below the water line is considered. The battery action that causes the metal on boats to corrode is discussed. The *grounding* methods used in the Pacific Intertie are unique. Solar winds can disrupt power distribution and damage oil pipelines.

5.1 INTRODUCTION

Our society requires the wide use of utility power. The environment we work in is full of power transmission lines. This same environment contains the radiation from many distant transmitters.

Power distribution presents different problems on a spacecraft, in an automobile, in an aircraft, or in a laboratory. In this chapter, we will

Grounding and Shielding: Circuits and Interference, Sixth Edition. Ralph Morrison.
© 2016 John Wiley & Sons, Inc. Published 2016 by John Wiley & Sons, Inc.

look at electrical power for facilities. Many of the interference processes that are discussed in this chapter are applicable to every power-operated system.

5.2 HISTORY

When utility electric power was first introduced, there were few rules and regulations. There were frequent electrical fires and people received electric shocks from poor wiring. During lightning storms, power wiring provided a path for lightning to enter facilities. Under pressure from insurance companies and the National Fire Protection Association, the NEC was initiated. This code provided standards that brought needed control to a growing power industry.

5.3 SEMANTICS

Key words used by a power engineer are defined by the NEC. These words must be clearly defined or the rules that follow can be twisted or misinterpreted. Circuit engineers do not have a controlling organization and their language keeps evolving. The word *ground* to a power engineer means a connection to earth or its equivalent. The word ground to a circuit engineer may mean a power common on the secondary of a transformer or a reference conductor in a floating circuit. In order to discuss the role of power wiring in electronics, we will use the power industry's definitions in this chapter. When we use these words to describe circuits, we will be less restrictive. Here are the definitions of key words and phrases used by the NEC. In this book, words that use the NEC definitions will be printed in italics.

1. *Ground* – A connection to earth or its equivalent
2. *Equipment ground* – All conductors that might contact power wiring. This includes conduit, equipment housings, racks, receptacle housings, trays, bare wires, and green wires. A green wire in power wiring is *equipment ground*. It does not carry power current. No other power conductor can be colored green.
3. *The grounded conductor* – A power conductor that carries current that is nominally at 0 V. It is usually colored white. It is earthed at the *service entrance*.
4. The *ungrounded conductor* – The power or "hot" conductor carrying voltage. The color code can be black or blue but never green or white.

5. *Neutral* – In three-phase power, the current-carrying conductor at 0 V. The *grounded conductor* is often called a *neutral* if it was derived from three-phase power.

6. *Isolated ground* – An *equipment ground* conductor that returns separately from a receptacle to a *service panel* or a *service entrance*. It is always earthed.

7. *Service entrance* – The power entry point into a facility.

8. *Grounding electrode system* – All of the interconnected non-power conductors in a facility including building steel, computer floors, equipment grounds, rebars, gas lines, and guy wires that could fault to a power conductor.

9. *Feeder circuit* – A group of conductors carrying power to *branch circuits* with a protecting breaker.

10. *Branch circuit* – A power circuit providing power to various loads with a protecting breaker.

11. *Separately derived power* – Power taken from an auxiliary power source or a distribution transformer where a new *neutral* is supplied. This *neutral* must be connected to the nearest point on the *grounding electrode system*.

12. *Listed equipment* – Electrical hardware that has been tested and approved for installation in facilities.

5.4 UTILITY POWER

When the power fails, our need is immediately apparent. Maintaining quality power is important. Local governments have accepted the NEC and now enforce its rules as the law. The code is modified regularly to accommodate new materials, methods, and procedures. The code represents good practice that has stood the test of time. The code is not a scholarly work with a rationale behind each rule. Often a practice is accepted because its rejection would represent a significant hardship for an industry.

We trust the electric appliances that we purchase. This safety is present because codes are followed and hardware is tested by various safety organizations before these items can be brought to market. In the United States, the phrase we all are familiar with is "UL Approved." Many design considerations go unnoticed. On a three-wire power plug, the round ground prong is longer than the two power connections. This forces the hardware to be *grounded* before power is connected thus avoiding a shock hazard.

The NEC requires that the metal housing of an electrical appliance be connected to *equipment ground*. The code permits the metal frame of an electric range to be connected to the *neutral* or *grounded conductor* when there is no *equipment ground* available.[1] This means there may be a small potential difference between the range and nearby *grounded* sinks or water faucets. This is the voltage drop in the *grounded conductor*.

The NEC now requires that power receptacles in hospitals to be tamper proof. Children are at risk when they try to poke things into the outlets. Tamper proof means that a spring-loaded shutter is pushed aside by the power plug.

The NEC disallows modifying *listed equipment*. Drilling holes in an outlet box is illegal. This means that all replacement parts will be compatible. Circuitry or extra wiring cannot be added to the power distribution system unless it is approved because this could compromise safety. There can be no surprises for a service electrician or for a contractor working on a facility.

Broadly, the NEC provides rules that keep electrical installations safe. If a fault condition occurs, no one will receive a shock and a breaker will immediately open the power circuit. The *listed* materials that are accepted for installation must provide service in all sorts of weather for an extended time period. Power wiring must be earthed so that lightning that couples to the power grid can find an earth path before it enters a residence or facility.

The utilities earth *grounded conductors* along their distribution path to limit lightning damage to distribution hardware. Their interest is to provide uninterrupted power to their customers. The rules for grounding in power generation and distribution are covered by separate codes. In a generating station, the fault currents can exceed 4000 A. The bonding of railings, metal floors, cable trays, generator frames, and so on becomes critical. The quality of the earth connection in a distribution station is also critical. High-voltage lines that leave a station can fault to earth and generators must be disconnected before there is damage. In the first cycles of the fault, a voltage drop between a handrail and a steel floor must not create a shock hazard or start a fire.

The NEC recognizes the need to protect *separately derived* power in a facility. The code permits these secondary sources of power to be

---

[1] Many residences built before WW2 did not use *equipment grounding conductors*. Outlet receptacles were all unpolarized two conductor types. It is illegal to replace a two-conductor receptacle with a three-conductor unit as this leaves the false impression that there is *ground*ing protection.

impedance grounded to limit the maximum fault current. This method of grounding is limited to voltage sources in the range 480–1000 V. Fault detectors are required to determine the position of the fault so that the right breakers are tripped when a fault occurs.

There are many ways to use the rules outlined by the NEC. Some approved practices are better at limiting electrical interference than others. The code makes no recommendations. It is for this reason that engineers must understand the mechanisms of interference. With this knowledge, decisions can be made about how hardware and facilities work together to control interference.

5.5 THE EARTH AS A CONDUCTOR

The earth is a complex conducting object. Typical resistance measurements between two points on the earth can vary from $10\,\Omega$ to megohms. The highest resistance areas might be blocks of granite, in the dry desert, or on a lava bed. The lowest resistance areas might be in damp soil or at the seashore. By burying copper conductors and treating the surrounding soil chemically, the resistance of an earth connection can be made as low as $1\,\Omega$ at frequencies below a few hundred cycles. Typical wet earth contact resistances are in the order of $10\,\Omega$. The contact resistance can be measured by applying a voltage between two earth connections. One half of the resistance observed can be attributed to each connection. Different spacings and contact areas will yield different answers.

N.B.

Low-impedance connections to earth require many parallel connections. In some areas, it may not be practical to make an earth connection.

Skin effect limits the penetration of power and lightning current into the earth. Lightning is characterized as being at 640 kHz. Obviously, power currents can go much deeper than lightning.

Lightning is an earth-seeking phenomena. To provide lightning protection, the NEC requires that the *grounded conductor* (*neutral* conductor) must be earthed at the *service entrance* to a facility. The resistance to earth should not exceed $25\,\Omega$. In poor soil situations, this resistance may not be attainable. In this case, two connections are sufficient to satisfy the code. When lightning strikes the power conductors, an earth connection provides a path for lightning outside of the facility. If the utility supplies power underground, the earth connection is still required. This practice forces all facilities to be alike with respect to earth *grounding*.

The power definition of an earth connection assumes that the measure is made at 60 Hz. The discussion in Chapter 2 makes it apparent that a long round conductor is not going to provide a low-impedance path between two ground points above a few hundred hertz.

Currents flowing in the earth and in the *grounding electrode system* of a facility are not easily controlled. Electrical interference is often blamed on this current flow. Any attempt to avoid this current flow by isolating a *grounding* area in a facility violates the code. The code allows only one *grounding electrode* system in a facility. The reason is very simple. The resistance between two 10-Ω earth connections could be 20 Ω. If a fault condition connects an *ungrounded* or "hot" conductor to the second *grounding electrode system*, a breaker may not trip. A 20-Ω load at 120 V is a current of only 6 A. In this situation, there could be a power voltage between conductors that would normally both be at *ground* potential. A fault condition under these conditions could thus cause a significant shock hazard that might go unnoticed.

The telephone industry uses the earth for ringing circuits. With the wide use of cable and rf transmission, few older telephone bells are in use. The use of a central office battery makes it possible to operate the phone system when the power grid goes down. Today batteries are located in most phones and ringing does not involve *grounded* conductors.

Electric trains use the rails as a return conductor. Some of this current return in the earth. The current pattern varies depending on the position of the train and soil conditions. In dc systems, there can be problems with electrolysis. The magnetic fields associated with this current flow can disturb a compass and in some cases cause errors in scientific measurement. BART in the San Francisco area uses a 1000-V dc system. A paddle makes a power connection at the side of each car. The return path for current is the rail system.

5.6 THE *NEUTRAL* CONNECTION TO EARTH

The NEC requires that the entry *neutral* or *grounded conductor* be earthed at the *service entrance*. This *neutral* may not be re*grounded* inside a facility. This insures that inside a facility, load currents do not flow in the *grounding electrode system*. The fault protection system in a facility relies on this restriction. Line filters can cause significant reactive current flow in the earth. See Section 5.13.

In many residences, the *service entrance-grounding conductor* is routed along an outside building wall. To protect this conductor from damage (lawn mowers) and to limit corrosion, this conductor is often carried to earth in a metal conduit. This concentric path is inductive and for this reason the conductor is bonded to the protective conduit at both ends. This means that lightning current will probably travel on the outside conduit surface not in the *grounding conductor*. The code still requires the central conductor even though it is redundant.

Where power is transported on utility poles, the *neutral* is multiply earthed along the transmission path. This provides a lightning path to earth and reduces the probability that the lightning current path will enter a distribution transformer. For wooden poles, this conductor travels down the pole protected by a wooden molding. The *neutral* conductor carries any unbalanced current in the three phases. The multiple grounding of the *neutral* means that some of this current can use the earth as a conductor. In areas with many facilities, earth currents can often flow in buried conductors such as building steel, gas lines, and fences.

The NEC requires that in a facility, an *equipment grounding conductor* must follow each power conductor and be connected at each metal power receptacle. This grid of conductors is a part of the *grounding electrode system*. For best practice in a facility, *grounded* conduit should surround what otherwise would be open power wiring. If a "hot" power conductor should fault to the *equipment ground*, the fault path must be must be low impedance to guarantee the fault current will be high enough to trip the breaker within a few power cycles. This is the reason why *equipment grounds* and power conductors must be run in parallel and close together. Any large loop in the fault path has inductance, and this can limit the fault current. Having the fault path nearby makes it easy to see that the path is present.

N.B.

In a residence, the code allows conduit to be used as the only *equipment grounding conductor* provided the conduit and hardware are *listed* and approved. In commercial facilities, it is a required practice to run a green *equipment grounding conductor* and still use the conduit as an *equipment ground*. A bare copper wire is also approved.

Flexible conduit is permitted by code. Its use is restricted to nonhazardous areas that are free from moisture. Liquid tight flexible tubing has an outer plastic covering and can be used in runs that are shorter than 6 ft. Flexible conduit cannot be buried in concrete under any condition. For runs greater than 6 ft, an *equipment grounding conductor* or green wire must be used in this conduit. The code frowns on flexible conduit that is less than 1 in. in diameter. Threading this conduit is not permitted. The code specifies the fittings that can be used with this type of conduit.

Equipment grounding conductors often contact earth at many points. Examples might be water pipes, boilers, motor housings, building steel, and metal siding. If power is routed on a metal surface, this surface should be connected to *equipment ground*. If there is a fault condition, this is the only way a breaker can interrupt the power and avoid a hazardous situation.

N.B.

There can only be one earth connection for a *neutral* power conductor in a facility.

N.B.

A *separately derived* source of power, such as a backup generator is earthed just as if it were a *service entrance* except there is no metering. There is still only one *grounding electrode system* in a facility even if there are multiple sources of power.

The *service entry* earth connection provides a path for lightning that uses the utility wiring and the earth as a transmission line. The *grounding conductor* in a residence often makes its earth connection to a nearby water pipe. This *grounding conductor* should take a straight line path from the *service entrance* to earth. A water pipe that is 10 ft away may not be effective.

5.7 GROUND POTENTIAL DIFFERENCES

A signal cable that is connected to a signal source at a remote point is associated with the ground potential at that remote point. When this unterminated but *grounded* cable is brought to a local circuit, the potential difference between ground points can be observed on an oscilloscope. A potential difference means that there are electromagnetic fields crossing the space between the cable and the earth. The voltage is simply the field intensity times the loop area formed by the cable, the earth, and the oscilloscope. If somehow the area could be made zero, there would be no voltage to measure. In the case of the earth, the currents associated with fields are distributed beneath the surface of the earth and there is no way to reduce this loop area to zero. If there is a continuous conductive plane involved, the loop areas and thus the voltage coupling can be controlled. The subject of ground planes is covered in Section 5.20.

In a facility, there are potential differences between racks and between pieces of hardware. These potential differences are again the result of electromagnetic fields in the area. There will be surface currents on all open conductors as a result of these fields. It is usually incorrect to use Ohm's law and assume a current flow based on a voltage and resistance measure. As an example, the noise voltage between two racks could measure 1.0 V. The dc resistance between bonded racks could be 1 mΩ. By Ohm's law, current flow would be 1000 A. This is obviously not the case. The same thing happens on the surface of a printed circuit board. A voltage between ground points represents field coupling to the area formed by the probe connection, not current flow.

Measuring resistances between ground points with an ohmmeter also has its complications. Digital or active meters may not function properly if there is power current flowing in the conductors. Often the contact resistance of the metering leads is greater than the resistance to be measured. One method of measurement is to use a four-terminal approach. A known current is applied to two outer contacts and the voltage difference between two inner points is observed. This method avoids the problem of contact resistance, but it is still just an approximation.

The resistance between two points depends on how the current distributes itself on the conductor. As an example, the resistance between two points on a square conducting plate will depend on the location and area of the contact points. Since the current does not flow in the corners, the shape or thickness of the corners cannot directly affect the resistance.

N.B.

The resistance between two points depends on how and where the current is inserted. The earth is not a component.

5.8 FIELD COUPLING TO POWER CONDUCTORS

Consider power transmission lines carried on power poles. A field moving in the direction of the transmission line can couple voltages to this line. There are two modes of coupling. The field that crosses the loop formed by the two conductors generates a normal-mode voltage. The field that crosses between the pair of conductors and the earth causes a common-mode voltage. This coupling is shown in Figure 5.1.

Normal-mode coupling adds to the transmitted signal. Common-mode coupling affects a group of conductors and appears as a ground potential difference. Ground potential differences can be attributed to current flow alone, but usually the voltages are the result of field coupling to the loop formed by the measurement leads. A common-mode signal is the average signal coupled to a group of conductors. Normal-mode

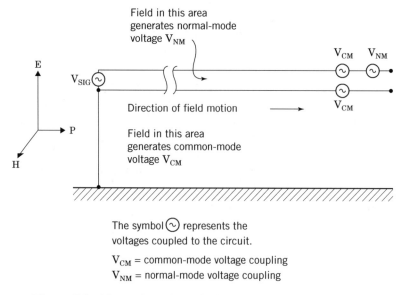

Figure 5.1   Normal-mode and common-mode field coupling

coupling affects each conductor pair separately. Normal mode is also called difference mode, differential mode, or transverse mode. Common mode is also called the longitudinal mode. In digital logic, a common-mode signal is called odd mode.

Common-mode fields couple to power conductors as well as signal conductors. For power, this coupling includes the *neutral* and the *equipment grounding conductor*. Coupled fields move energy in both directions. There are reflections that occur along the line as well as at the ends. These reflections do not affect the power transmission but they do add to the field energy that is brought to the hardware via the power connection. Power line filters can reflect common-mode and normal-mode interference at frequencies above about 100 kHz.

There are many ways for field energy to couple into hardware. Power line filters are only a part of the story. In Chapter 7, the coupling mechanisms are discussed in detail.

5.9 *NEUTRAL* CONDUCTORS

Utility power is generated three-phase because the power generated for a balanced load is constant over the entire cycle. Constant power generation means the torque on the rotor of a generator is a constant over each rotation. Heavy equipment such as large fans and industrial motors are connected directly to the three phases. For most low-power applications, power is taken from phase to *neutral*. The loads are arranged so that each phase supplies the same amount of power. If the loads were linear and balanced, the *neutral* currents would average zero throughout each cycle.

Many electronic devices use a rectifier system that store energy in filter capacitors. The capacitors demand current near the peak of voltage. These peaks of current occur at different times for each phase with the result the *neutral* currents cannot balance to zero. This *neutral* current is rich in harmonics, and it flows in the reactance (series inductance plus the series resistance) of the *neutral* conductor. The resulting voltage drop in the *neutral* conductor can be seen as a potential difference between *the equipment ground* and the *grounded conductor*. This voltage is in series with leakage capacitance in the power transformer and causes current to flow in the input common conductors. See Figure 5.2.

If a distribution transformer handles loads with high harmonic content, the magnetic leakage field around that transformer must carry this

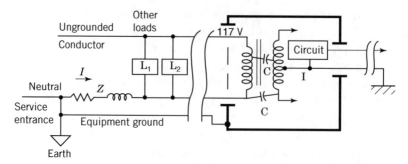

Figure 5.2   The coupling in a transformer resulting from neutral voltage drop

same harmonic content. This means the nearby B field is rich in harmonics. Any high-frequency content in the field can more easily couple to nearby conducting loops. If the building steel forms a nearby loop, these induced currents can circulate in an entire facility. If a separate steel structure is used to mount the transformer, it is good practice to break up the loops of steel by using insulators.

In three-phase power, if the harmonic content is high, the harmonic *neutral* current can be greater than the line current. *Neutral* conductors that carry high harmonic current can overheat. If the power is carried in a delta connection, then there is no *neutral* conductor to consider. This approach is more expensive as it requires an added *separately derived* source of power. This added transformer must have a *grounded neutral*. In effect, a *separately derived* source of power is like a new *service entrance*. Of course, the loads on this secondary can again generate a *neutral* voltage drop.

5.10 *k* FACTOR IN TRANSFORMERS

Electronic loads are typically very nonlinear. Rectifier systems require large amounts of current at the peak of voltage. This nonlinear current can be described in terms of harmonics of the fundamental. The effect on transformers is to increase eddy current losses in the transformer iron and increase heating losses in the wiring. Transformers used to supply electronic loads must be rated to handle the harmonic content, or they will overheat. This problem occurs often enough that a special factor is used in specifying these transformers. This factor is known as the "*k*" rating. A high *k* rating requires transformers be built with a higher grade of transformer steel and with larger wire sizes. To limit skin effect, the conductors may be laminated.

There are several ways to calculate the k rating. The simplest method is to use Equation 3.1. In this equation, n is the harmonic number, I_n is the current at this harmonic, and I_T is the total load current.

$$k = \Sigma n^2 \left(\frac{I_n}{I_T} \right)^2 . \qquad (3.1)$$

If there is no harmonic content, the k factor is unity. Harmonics out to the 25th should be considered in calculating k. Typical k factors for electronic loads are around 8 while factors above 30 are rare.

5.11 POWER FACTOR CORRECTION

Utilities are faced with a problem in delivering power to industry where large motors and fans are operated. These loads are quite inductive and demand reactive current. The utilities are not paid for supplying reactive energy even though their generators and transmission lines are tied up. To solve this problem, the utilities can add capacitors to the line to correct for this reactive current flow. They can also use rotating hardware to supply reactive current. This correcting hardware should ideally be located near the load. Since the reactive load changes throughout the day, the amount of correction must be variable. Switching capacitors onto a utility line introduces transients. Inrush currents that last a few milliseconds can be limited by using series surge resistors that are shorted out in a few cycles. Note that the switching involves the three phases with zero crossings occurring at different times. Online rotating hardware correction introduces no transients.

5.12 UNGROUNDED POWER

There are a few installations, where un*grounded* power is required. An example might be power used to heat a large crucible. If there is a fault and power is interrupted the crucible might be lost. The procedure is to detect the fault and turn the power off when the crucible is empty. Facilities that use ungrounded power must have personnel on duty that are properly trained to handle any fault condition that might occur.

The power supplied on shipboard in the navy is floating. During battle, a ground fault should not disable critical functions. The issue of electrolysis and moisture must also be considered. Floating power is electrically noisy. When power is switched on or off to a load, the distributed capacitances in the entire system must store different amounts of energy. The transients that result enter every piece of equipment through the device power transformers. One approach used to avoid this interference is to add a distribution transformer that supplies *grounded* power to sensitive electronic hardware.

5.13 A REQUEST FOR POWER

What happens when a power switch is closed in a facility? The answer may surprise you.

Before the moment of closure, the electric field between the *grounded and ungrounded conductors* stores electrostatic field energy. At the moment of closure, waves at half voltage start moving to the load and to the voltage source. The current in the waves depends on the characteristic impedance of the line. When the wave reaches the load, a reflection takes place based on the impedance of the load. Meanwhile, a wave of half that voltage propagates back toward the *service entrance*. The wave progresses until it reaches a branch connection. At this point, there is another reflection, and the wave propagates along two paths. The process of branching and reflecting soon fills the entire wiring grid in the facility with electrical activity. The electrostatic energy stored in the wiring begins to move toward the load to supply it with power.

The electrical activity immediately after the switch closure involves numerous transmissions and reflections that combine to propagate waves that eventually reach the power generator that is miles away. The transient waves at the *service entrance* no longer have a steep wave front, and the waves lose amplitude. It is interesting to measure the transient voltages on a power line at different distances from a switch closure. Near the switch, the voltage drops to half value for a few microseconds. At a nearby panel, the voltage transient might be a volt lasting a millisecond. At the service entrance, the transient is hardly visible. Waves must eventually reach the power generator, so that it can adjust its output voltage to accommodate the new demand.

Consider the electrostatic energy that is available in a facility. Assume a capacitance of 50 pF per foot of wiring. In 1 μs, a wave will propagate out a distance of 500 ft. The network within this radius might consist of 3000 ft of wiring. The wiring capacitance might be 0.15 μF. The energy stored in this capacitance when the voltage reaches 170 V peak is about 10^{-3} J. If this energy were dissipated in 1 μs, the power level would be 1000 W. If the load request is 100 W, the voltage at a distance of 500 ft would sag by about 5%.

The energy that bounces around the power wiring circuitry moves in the space between power conductors. Some of this energy is confined to the space inside metal conduit. Some of this energy adds to the ambient field in a facility.

Very rapid voltage changes on the power line have little effect on the rectified and filtered supplies inside the hardware. There are still several coupling mechanisms that allow fields to enter the hardware that involve the power transformers. If the power filters are not installed correctly, then this is an entry point for interference. These entry mechanisms are discussed in Chapter 7.

5.14 EARTH POWER CURRENTS

Power wiring that distributes power outside of facilities is often four-wire. This means the *neutral* conductor is carried along with the three phases. The *neutral* is earthed along the distribution path as a protection against lightning. The amount of *neutral* current that flows depends on load unbalance and on harmonic content. The *neutral* current that flows divides between the earth and the *neutral conductor*. The current density between earth points is low, where the current spreads out in the earth. A wide current path means a low-impedance path. A typical earth resistivity might be $1000\,\Omega$-cm. Using this figure, the resistance across a cube of soil $10\,m$ on a side is only $1.0\,\Omega$.

In areas using electrical power, there are often metal objects buried in the earth that reduce the impedance between points in the earth. Gas lines, metal fences, and building steel are good examples of buried objects. Building steel is earthed at many points for lightning protection. *Neutral* current flow will follow these conductors as they provide a lower impedance than the earth. Note that the *neutral* current flowing in building steel could be associated with utility power flowing to a nearby facility.

5.15 LINE FILTERS

Electronic hardware is often supplied with line filters. These filters place capacitors between the power leads and *equipment ground*. *Equipment grounds* form a grid that can be earthed at many points. This means that reactive filter currents can use parallel paths to return to the *neutral* at the *service entrance*.

One function of a line filter is to keep interfering fields from entering the hardware. The filter works by providing a high-impedance path into the hardware and a low-impedance path around the hardware for high-frequency energy. Filter capacitors circulate interference currents as well as 60-Hz currents in the *equipment ground* grid in a facility. All the return paths lead to the *service entrance*, where *equipment grounds* connect to the *neutral* conductor and earth. If there are many parallel paths for current flow, the field intensity will be low. The number of filters is proportional to the number of hardware items involved. In large systems, the reactive currents returning to the *neutral* tie at the *service entrance* can be amperes. Obviously, filter capacitors will carry utility current as well as interference.

Line filters can provide a local source of immediate energy for hardware. If the hardware requires energy in a step manner, the capacitors in the local power filter can supply this leading edge energy. This stops a high-frequency request from propagating a steep wave front out into the facility. This energy is not available if there is a series inductor on the load side of the filter.

N.B.

The 60-Hz current in a filter capacitor is reactive, which means it is not in phase with the voltage. This fact can be used in troubleshooting to identify the source of some interference. Power line filters are discussed again in Chapter 7.

5.16 ISOLATED GROUNDS

The NEC allows for *equipment grounds* to be "isolated" from each other and only *grounded* at the *service entrance*. The concern is that interference can cross couple between hardware if they share the same conductors. The assumption that is made is that by providing separate *equipment ground*ing conductors that go to the *service entrance*, this will limit any cross coupling. This reasoning is flawed as the line filter in each piece of hardware connects capacitors from both sides of the line to this *equipment ground*. The inductance of a long *equipment ground* conductor can make the line filters ineffective. If there are signal paths that interconnect the hardware, power line filter currents will flow in this cross connection. This can couple interference into a signal process depending on the nature of the interconnecting cable. This problem is shown in Figure 5.3.

In stand-alone hardware, power connections are often made to separate power receptacles. In the *isolated grounding* arrangement, the loop

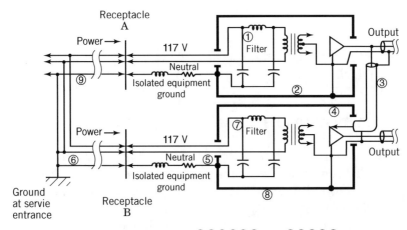

Filter current flows in path ①②③④⑤⑥ and ⑦⑧③②⑨.
Path ③ is a signal conductor.

Figure 5.3   Filter current flowing in signal interconnections

formed by the *equipment grounds* can be large. Assume a signal path between the two pieces of hardware closes this conducting loop. Transient fields from nearby current interruptions such as in a motor can couple to this loop and induce common-mode voltages. At a signal interface such as in Figure 5.3, these voltages can be great enough to damage hardware. The standard *equipment ground* configuration limits this loop area. The result is a more reliable installation.

The rationale for using *isolated grounds* is to keep the contaminating current from one piece of hardware from flowing in the *equipment ground conductors* of other hardware. Unfortunately, the current flowing in an *isolated ground conductor* can create fields that can fill an entire facility. The problem is not the current flow but the fields that are created by the current flow. In nearby conductors, the fields imply surface currents. If every equipment ground has a long path, the fields in the facility are much greater. The fields associated with a grid of *equipment grounding conductors* are less intense. If the current spreads out on a ground plane, the field intensity can be further reduced.

N.B.

A grid of *equipment ground conductors* behaves like a pseudo ground plane.

> **N.B.**
>
> The NEC has banned the use of *isolated grounds* in hospitals and medical facilities.

5.17 FACILITY GROUNDS – SOME MORE HISTORY

The ideas behind a central facility ground probably started in the early days of electronics. In designing hardware, it was found that using a water pipe ground reduced hum coupling in analog amplifiers. There was evidence to suggest that a better earth ground reduced the interference further. The exact reasons were not debated because the solution seemed obvious. The search for the best ground had begun.

It was common practice in circuit design to use a single-point grounding scheme. The idea was to control cross coupling due to the flow of current in common conductors. The common conductors included power supply conductors, input signal leads, output common leads, shield connections, and the equipment ground. These conductors were stacked up on a common grounding stud mounted to the chassis. The order of connection was carefully considered. This arrangement of grounds is called a star connection. As an example, a star connection insures that output current does not flow in an input common conductor. Also, current flowing in an *equipment ground* will flow directly to the hardware chassis without flowing in a signal conductor. A star configuration works fine if the dimensions are under 100 cm and the frequencies of interest are under 20 kHz. In today's digital world, this practice cannot function. It cannot function at the circuit card level or at the facility level for *equipment grounds*.

> **N.B.**
>
> The use of a star connection has its place. The connection of the *neutral*, the *equipment grounds*, and the earth at *the service entrance* is a star connection.

Circuit designers like to think of a ground as being able to "absorb" noise. After all, capacitors connecting circuit points to this ground seem to "drain" the noise away. It seems obvious that the earth is the ultimate current sink. Capacitors have this ability to "bypass" or "sink" noise. These ideas have little scientific basis.

In the 1950s, the electronics industry received a big boost from aerospace and military activity. Many large electronic facilities were built in this period. The idea of a single-point ground was extended from hardware design to an entire facility. A typical facility might contain hundreds of electronic devices. If a good ground made an instrument quiet, then a "really good earth ground" would make a facility really quiet. As a result, many facilities were built with a very expensive and extensive grounding schemes. The star connection idea was extended to a grounding well where an earth connection was made using large amounts of copper and chemically treated soil. All signal shields were assembled at several nodes and carried as one large conductor to this earth ground. A similar treatment was made for all *equipment grounds* and signal commons. In these schemes, the power was still supplied to a *grounded service entrance*. This earth connection was separate from the ground well.

Bringing all the *equipment grounds* via a collecting node to this one star *ground* point added significant loop areas to the fault protection path. It was argued that this departure from the code was needed to provide for a quiet facility. This practice added a loop area to every return path for filter current. This added to the interference fields in a facility rather than making it quieter.

The premise that noise currents flow into earth and never return is of course not good physics. I call this the "sump theory of electronics." Circuit theory requires a return path for currents that go the earth. Where the return paths are, nobody knows. Engineers seemed to brush this issue aside thinking, "Somebody knows more than I do." Once a complex grounding scheme is installed, it is difficult to disable it. It is also very difficult to experiment with an entire facility to prove or disprove performance. It is also very difficult to argue with a bureaucracy.

N.B.

A facility is not a circuit. It is a complex of conductors, where fields associated various voltages and currents follow the paths that store the least amount of field energy.

A central grounding conductor that "carries all interference current to earth" is a giant upside down antenna. The interference fields around this ground are propagated through an entire facility. In this approach, it is difficult to stop radiation.

5.18 GROUND PLANES IN FACILITIES

A ground plane is a conducting surface. The size of a ground plane depends on application. For a printed circuit board, the size of the plane might be as small as a square centimeter. A large printed circuit board ground plane might be 12×18 in. In an electronics facility, a constructed ground plane might be the size of a room. Ground planes include the side panels in racks, a strip of aluminum foil, or for that matter the earth under a building. A metal box can be considered a ground plane although the conducting surface is not flat. If the conducting surface extends from a floor on to a wall it is still a ground plane. It is obvious that the word plane had lost its "flat" character.

In Chapter 3, the printed circuit board ground plane was used as a return path for logic and power currents. Stated another way, the fields that carry logic or signal information are carried between traces and the ground plane. If a conducting surface is flexible and it turns a right angle, it still functions as a ground plane.

N.B.

It is not practical to use the ground plane in a facility to provide a return path for logic or signal information.

Ground planes have a place in facility design. Their presence is not a guarantee that interference will be limited. Like any other tool, the benefits must be weighed against cost. The size and nature of facilities change as the technology shifts. What took a room full of hardware now requires one rack of equipment. What required extensive cables now uses fiber optics. What remains constant is the character of utility power and the nature of lightning and electrostatic discharge (ESD). The discussion that follows is not a recommendation. It is a review of how ground planes can be used in facilities.

A facility ground plane is often a grid of stringers supported on stanchions. The spacing between stringers might be 30 in. The stanchions are usually about 3 ft high.

There are several benefits to having a raised floor. Cables can be routed between racks under the floor and this allows for a clean looking installation. The space under the floor can form a plenum chamber that can be used to supply cooling air for the electronics. The floor tiles that are supported between the stringers are made slightly conductive to drain charges from humans that walk on the floor. This controls ESD

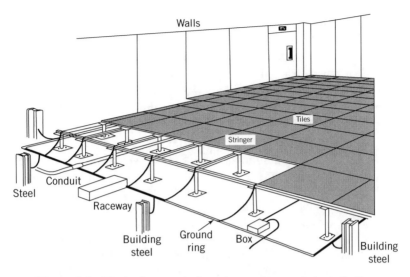

Figure 5.4 Typical ground plane in an electronic installation

that can be very destructive. A typical human body has a capacitance to the floor of about 300 pF. A time constant of a 10th second means the floor tiles should provide a resistive path of about 300 MΩ to allow an accumulated charge on a human to dissipate quickly. These conductive tiles usually have a resistivity of about 10^7 Ω-cm.

If the grid of stringers is connected to building steel around the perimeter of the room, then during periods of lightning activity, there will be limited potential differences between racks of equipment. This structure is shown in Figure 5.4.

A facility ground plane does not attenuate or eliminate electromagnetic fields. The only role a conductive plane can have is to reflect an arriving wave that has an E field that has a component that is tangent to the plane. Waves with a vertical E field can propagate along the surface of the plane without attenuation. The associated H field that is parallel to the conducting surface simply causes a surface current to flow.

To make the stringer ground plane look as good as a sheet of copper, the resistive connections between the stringers should be microhms. This is accomplished by manufacturing the stringers with plated surfaces and using spring-type washers to bolt the stringers together. This way the contact area is under constant pressure. The intent is to maintain a good connection over time for a range of floor loads and temperature changes.

The stringer ground plane should be connected to each rack of electronics, which in turn is connected to *equipment ground* for every piece of equipment. The line filters in equipment have shunt capacitors that connect to equipment ground. The stringer system connects to this equipment ground and provides multiple paths for these power filter currents. For this reason, the stringer system is effective in limiting the intensity of interference fields associated with this current flow.

Each tile that fits between the stringers must make contact with the stringers around the perimeter of the tile. When tiles are removed and then replaced, the hardware that provides this connection must be reinstalled or the protection against charge buildup will not be present.

N.B.

The best way to control ESD is to control humidity. If the humidity is above 30%, there is little chance of charge buildup. The best way to control humidity is in a central air-conditioning unit. Added humidifiers often leave areas without adequate protection.

N.B.

To avoid generating ESD, rotating floor polishers should not be used on floors near electronic hardware.

The stringers of a ground plane act like a wire mesh except that the openings are much larger. For a wire mesh to be effective, it must be bonded to an enclosure around its perimeter. See Section 7.10. In the case of a ground plane made from stringers, there are no controlled conductors around the perimeter to use for aperture closure. For this reason, external fields can be present on both sides of the ground plane.

For a ground plane to be effective in a facility, all equipment racks should be bonded to the ground plane at their base. In effect, the racks should be extensions of the ground plane. Of even more significance is the routing of connecting cables. The usual procedure is to drop the cables to the concrete floor under the racks. The loop areas between the cables and the stringer ground plane can be large. These loop areas allow common-mode coupling for fields that exist under the stringer floor. Hanging cables on the stringers can reduce this coupling although it is rarely done.

Racks that are bolted together should use contact washers under pressure against plated surfaces. Vibration should not loosen these connections.

5.19 OTHER GROUND PLANES

The rebars in a concrete slab might be considered a ground plane. Rebars are buried in the concrete and multiple surface connections are not provided. Because the bars are not welded at every crossing they do not constitute a proper grid structure. The presence of this steel should have little impact on the electrical performance of a facility. Rebars are simply more conductive materials such as water pipes, gas lines, steel beams, or metal trays. These conductors modify electromagnetic fields, but they are not controlled surfaces.

A conducting surface can be added to the floor of connected racks to form a ground plane. This added surface must be bonded across its width at the ends to the rack framework to be effective. This added conductor should not concentrate current flow at single points. A ground plane only has meaning if it controls or limits field coupling.

5.20 GROUND AT REMOTE SITES

Remote sites often house electronics in a trailer. If utility power is provided to the site, there can be a safety issue. If the site is on the desert floor, then an adequate local earth connection may not be available. The vehicle frame becomes the *grounding electrode system*. All metal objects external to the vehicle should be connected to the vehicle. If this is not done, there is a possible shock hazard. If there is a power fault or if there is a lightning strike nearby, there can be a potential difference between the vehicle and these conductive external objects. For example, a metal floor mat and a nearby metal fence should be bonded to the vehicle chassis. If this bond is not present, then an electrical event can be dangerous.

5.21 EXTENDING GROUND PLANES

If it is required to extend a ground plane from one room to another, it is sufficient to make a series of spaced connections through the separating wall. A typical solution would be to connect the stringers together using #10 wire on 15″ centers across the width of the room. The conductors might be welded or bolted to a prepared surface on the stringers.

If it is required to extend a ground plane to a second floor, the planes must connect by an extension up one wall. In this case, cables routed between floors must cross to the other side of the plane. This crossing should pose no problem. The plane on the wall can be made of a wire mesh as long as the conductors are bonded at every intersection. The mesh should be welded or bolted to the stringer system along the width of the room. The holes in the mesh allow fields to cross through to the other side. Cables must also pass through this mesh.

It is not practical to attempt to use one ground plane that extends between buildings. An accepted compromise is to place cables inside a large conducting conduit and bond the conduit to the ground planes on each end. The fields associated with the lightning will probably stay on the outside of the conduit. This reduces the risk of lightning following one of the cables into the hardware. Optical fibers or rf links are an effective way to transport signals between buildings without having to consider ground potential differences.

5.22 LIGHTNING

The electrical activity around the earth is complex. A lot is known, but the entire story is still being studied. Here are a few of the facts that are known. The sun bombards the earth with a steady stream of electrons and protons that are mainly deflected by the earth's magnetic field. Cosmic rays are not deflected and strip charges off of atoms. The result is a region in the space at the top of our atmosphere that is called the ionosphere. This is at an altitude of about 50,000 km. The ions in this region are free to move long distances and this maintains a nearly constant potential to earth of about 400,000 V. The voltage gradient at this altitude is small.

At the surface of the earth where the atmosphere is dense, the voltage gradient increases to about 100 V/m. If a large insulated conducting surface is carefully suspended in a dry area, it will eventually collect a charge based on its capacitance and the voltage gradient. This tells us that there is a stream of charged particles in constant motion all over the earth. For a metal plate, this arriving current is only micro-micro amperes. For the entire earth, the current level is estimated at 1800 A. This current times 400,000 V is 720 MW. This is just one small component of the energy arriving at the earth from the sun.

The voltage gradient near the earth has been known for some time. It is interesting to note that this gradient is minimum at about 4:00 AM Greenwich Mean Time. It turns out that lightning activity around the world peaks in Africa at this hour and this explains the daily fluctuation in this voltage gradient. It is estimated that there are over 1000 lightning strikes to the earth every second.

When it rains, water droplets strip electrons from air molecules and carry this charge to the earth. In areas of weather, the local electric field intensity can increase and the field strength around sharp objects can be quite high. If the air ionizes around this object, the ionization path extends the sharp point upward. The voltage gradient between the clouds and the tip of the ionization path increases and this accelerates the ionization process resulting in a narrow conducting path from the earth to a cloud of ions. At this time, there is an avalanche of charge that flows down the ionized path. The electric field in the ion cloud stores the electric field energy that uses the ionized path as a conductor. It is as if a long conductor were discharging an immense capacitor storing charge.

The volume in space that is discharged extends about 150 m around the ionized path.

At the speed of light, it takes about a half microsecond for this field to collapse. The pulse of current that flows can vary from a few thousand amperes to a maximum of 100,000 A. The first pulse of current widens the ionized path. After this current pulse stops, the field pattern readjusts in the cloud and a second and third pulse of current can follow using this same path. When the path breaks up or the voltage gradient is too low, the lightning activity stops. A single pulse lasts a few microseconds.

N.B.

Lightning is one of the mechanisms that keep the charge in balance in the atmosphere all over the earth.

The voltage gradient required to ionize air is enormous. For the path lengths involved, the potential differences in a cloud can be 10,000,000 V. The charge that flows can be 20 coulombs. This means that the power level in a pulse that lasts a few microseconds approaches a gigawatt.

5.23 LIGHTNING AND FACILITIES

When lightning strikes the earth, the current spreads out radially. Because of skin effect, the electromagnetic field cannot penetrate very far into the earth. The voltage gradient near the strike can be great enough to electrocute cattle standing in a wet field.

Utilities should bring services into a structure at the same point. Examples might be power, cable, and telephone lines. If separate grounding points are used, a nearby lightning strike can cause large ground potential differences between these utilities. If these spaced and grounded conductors are brought together into the same hardware, there can be damage to the hardware.

An example will illustrate the problem. A contractor installed motion detectors and a dial-up reporting system in barracks at an unused army base. The telephone lines and the utility lines were grounded at opposite ends of the building. The first lightning storm destroyed the installed electronics.

Lightning and circuit theory have little in common. The electromagnetic field around the lightning pulse is complex. It is this rapidly changing field that controls the path the lightning takes. Our best tool to describe this current path is in terms of inductance. At the moment the main pulse reaches the building, the voltage to earth is near 0 V. The energy that flows through the building can be thought of as a wave in a coaxial transmission line. The lightning path is the center conductor. The outer conductor (return path) is the displacement current flowing in the surrounding space. The displacement current is the rapidly changing E field.

The voltages in various conductor geometries can be calculated assuming a rate of change of current in an inductance. The pulse of current rises in about $0.5\,\mu s$. A 10-in.-long conductor has an inductance of $1\,\mu H$. A 50,000-A pulse develops a voltage of 100,000 V in this inductance. Near a building, the voltage pulse could easily reach a million volts per floor.

When voltages of this magnitude are involved, air can ionize and conduct. When lightning hits a steel structure such as a building under construction, one can see lightning flash between girders rather than flow in a horizontal beam out to a vertical beam. The reason is simple. A right angle path is inductive and the resulting voltage will ionize the air. The air breakdown starts at the beam intersection and moves out into the surrounding space. When lightning hits the roof of a building it should be directed by conductors to the walls of the facility and to earth on parallel "down" conductors. The added lightning paths should be lower impedance than air ducts, plumbing vents, or antennas.

Where possible, down conductors should provide straight paths without bends or turns. If there must be turns, they should have a controlled radius. A wide strip of metal is a far better down conductor than a long round conductor. In buildings with sheet metal sidings, the sidings provide a far lower impedance path than a group of down conductors. In this case, added down conductors may actually be redundant. If sheet metal is used, it should be earthed at multiple points along the bottom edge to limit arcing to earth near the ground floor of the facility.

Lightning can develop a high induction field near a down conductor. In a building with a steel framework, electronics should be positioned away from this steel to avoid possible damage. To illustrate this problem, the H field at a distance of 1 m from a 50,000-A pulse is almost 8000 A/m. The B field is $\mu_0 H$, which equals 0.01 T. The flux in a conducting loop of 0.01 m^2 is 10^{-4} lines. If the flux rises in 0.5 μs the induced voltage is 200 V. This changing flux coupled to a small loop can destroy electronics. The induced voltage near sheet metal siding carrying this same current is probably reduced by a factor of 10.

N.B.

Lightning current does not need to flow in a circuit to do damage. A nearby current pulse can induce significant voltage into a conductive loop to damage circuitry.

Lightning currents should not be allowed to concentrate, as this is where the most heating will take place. In one case, lighting hit a metal tower mounted to a rock outcropping. The path to earth concentrated on one of the tower legs melting the metal at the foot of the tower. The tower collapsed on a fuel tank and the resulting fire destroyed a remote facility.

When the statue of liberty was refurbished, there was an examination made for lightning damage. There were numerous pit parks where there had been strikes. There was no damage as current was immediately dispersed over the large surface. In the tower example above, if the current had been dispersed at the base into a metal mesh, there probably would have been no damage.

5.24 LIGHTNING PROTECTION FOR BOATS AND SHIPS

In areas where there are frequent thunderstorms, watercraft are often struck by lightning. The presence of a mast or antenna provides an attraction point for the strike. Small craft are particularly vulnerable because they are usually not metal structures. Lightning is an earth-seeking phenomena, which means in this case the surface of the ocean. In a metal craft, the current flow is dispersed and flows into the water over a wide area. In a small craft, the entry points into the water are limited.

N.B.

In the literature the ocean surface is referred to as ground.

The complication that is often missed is that current cannot flow from a keel or a submerged propeller shaft into the water. This is because the ocean is a conductor and fields are attenuated in conductors. Remember it takes fields to have current flow. The metal hull works because current can enter the ocean at the ocean surface over a wide surface area. In a small craft, there may not be a simple conducting path to the water surface with the result that there will be arcing between conducting objects on the craft. In this situation a human being is a conducting object.

The energy involved in lightning activity is in the electric field in the cloud. The discharge of this energy involves light, sound, and heat. The lightning current that flows in the ocean is limited by skin effect to the top few feet of water. The first consideration is to keep lightning from entering an antenna and destroying electronics. A pointed grounding conductor (also called a lightning rod, air terminal, or down conductor) mounted higher than the antenna can be provided. The antenna signal can be carried in coax that can be strapped to the air terminal. A curve in the coaxial path will insure that the lightning current will follow the straight path. The next problem is to provide conducting paths from any air terminals to the ocean surface. A group of these down conductors forms what is called a Faraday cage. These conductors act like a protecting umbrella that limits the penetration of fields. In this application the lightning current follows down conductors in its path to the ocean surface.

To be effective the down conductors should be terminated on "grounding plates." The preferred position of the plates is at the water line as lightning current (fields) cannot penetrate much more than a meter into the ocean water's surface.

If down conductors are not provided and there is a lightning strike, the question is how does the energy reach the water. Arcing will generally occur between available conductors to reach the water. Conductors can include electrical wiring, conduit, electrical hardware, or plumbing. The path can literally blow a hole in an insulating surface including the hull to get to the water.

5.25 GROUNDING OF BOATS AND SHIPS AT DOCK

We started this book by discussing the electric and magnetic fields associated with charges. These charges are the electrons that surround the nucleus of atoms. In conductors, the number of electrons involved in electrical activity is an extremely small fraction of the number available. It is interesting to note that the properties of metals, gases, and dielectrics are governed by how the fields from these electrons interact at an atomic level. The fact that conductors maintain their physical properties at very high current levels is another way of saying that hardly any of the available electrons are involved in current flow.

Chemistry is the study of how atoms interact. Basically this interaction involves sharing electrons in complex molecular structures. Most of the metals that we take for granted are rarely found in nature in a pure state. A good example of this is the alloys of iron that we call steel. Nature gives us iron ore, which is ferric oxide. We must do a lot of work to separate iron from oxygen and then add various other elements to get a good structural material. The same is true for aluminum, which is obtained from bauxite ore by electrolysis. What is important to realize is that nature is constantly trying to combine metals with other atoms as this represents a lower energy state. For iron this atom is oxygen and this molecule is plain old rust.

A lot of chemistry can occur when a conducting liquid such as seawater comes in contact with a metal. When salt dissolves in water, the ions of sodium and chlorine atoms are not bound together as they would be in a crystal. These ions are separated in the sense that they can support current flow in the liquid. Ionic current flow is the basis of plating.

Consider two ships in contact with seawater; one has an aluminum hull and the other a steel hull. This forms a battery and a potential difference can be measured between the metals. If there is a conductive path between the hulls such as *equipment grounding conductors* from shore power, current will flow from the battery through the seawater. The chemical action will oxidize the aluminum hull. It takes three electrons to bond two atoms of oxygen to form one aluminum oxide molecule. This is correctly called galvanic corrosion.

There are two methods available for limiting this corrosion. The first is the use of galvanic isolators. These isolators block small differences of potential from generating current flow in *equipment grounding conductors*. A pair of silicon diodes wired back to back will block about 0.6 V. If placed in series with an *equipment ground* connection, galvanic action is stopped but fault protection will still be provided. The second method involves the use of a sacrificial anode. The material that is preferred is zinc as it is high in the electromotive series. The zinc is oxidized rather than the aluminum. The zinc must make a good electrical connection to the aluminum hull below the water line. If large surfaces are involved, then several zinc anodes may be required. At dc, currents do flow beneath the water surface.

Protecting boats and ships from galvanic corrosion and corrosion from stray dc current is well understood. There are many different conditions to consider. Often craft are built using combinations of aluminum, stainless steel, and iron. There need not be a grounding conductor for corrosion to occur. There are a few subtle problems involving crevices, cracks, plastic insulators, and tightly wound ropes where air cannot circulate. These weak points can allow moisture to oxidize metals causing damage. There are even chemical processes that generate acids in the water that can do damage to wooden hulls.

5.26 AIRCRAFT GROUNDING (FUELING)

Air moving over the skin of a metal aircraft can easily cause a buildup of charge in flight. On landing, this charge will eventually bleed off to ground. The bleed rate depends on the humidity and the condition of the tires. It is standard practice to ground an aircraft before fueling operations begin. Grounding here means an electrical connection between the fuel truck, the aircraft, and earth. This is done before the fuel tank cap is opened. Any arcing (ESD) from the aircraft to the fuel line could result in an explosion and fire. Fueling personnel are required to wear special shoes and clothing to avoid the risk of ESD.

Aircraft are often struck by lightning in flight. The lightning uses the aircraft as a part of lightning path. As mentioned earlier, lightning usually consists of multiple pulses. The distance traveled by the aircraft between pulses will show up as a series of inline pit marks. The fuel caps on aircraft are built to keep surface current from getting under the caps where there is fuel vapor.

5.27 GROUND FAULT INTERRUPTION (GFI)

The NEC now requires that ground fault interruption (GFI) power receptacles be used in kitchens, bathrooms as well as outdoors where electrical appliances or devices might be used. These receptacles interrupt the power if 5 mA or more of current does not return in the *grounded conductor*. A typical path might involve an appliance where the metal housing has developed a leakage path to the *ungrounded* (hot) conductor. The user gets a shock if he makes an electrical connection between a metal sink and the appliance. Current in the 5 mA range can be lethal.

The circuitry in a GFI is complex. The *grounded and ungrounded conductors* are threaded through a small magnetic toroid. If the current flow is exactly equal and opposite, the net ampere-turns in the magnetic path is zero. If there is a current imbalance, a coil of a few hundred turns on the toroid generates a signal of several volts. This signal operates a circuit that then operates a relay that disconnects the power from the outlet. The relay is latched "on" and can only be reset mechanically. Much of this circuitry is incorporated into a semiconductor component. Without this miniaturization the product would not be practical.

N.B.

When there are multiple outlets on one branch circuit, only one outlet needs to be a GFI. The outlets that are chained beyond the GFI will be automatically protected.

If an outlet is inactive and the circuit breaker is normal, the problem may be that a GFI has been tripped.

5.28 ISOLATION TRANSFORMERS

Isolation transformers are one type of *separately derived power*. They are often used in facilities that operate heavy equipment such as welders, intermittent electric motors, hoists, and punch presses. These loads can create interference problems for laboratories or testing facilities depending on how the facility is designed.

An isolation transformer provides a new *grounded or neutral conductor*. This eliminates a big part of the interference problem. There are two more ways this added transformer can reduce interference. Electrostatic shields can be added between the primary and secondary transformer coils and a line filters can be placed on the transformer primary. A transformer with shields, filters, and a breaker is often called a computer power center. The use of shields is shown in Figure 5.5.

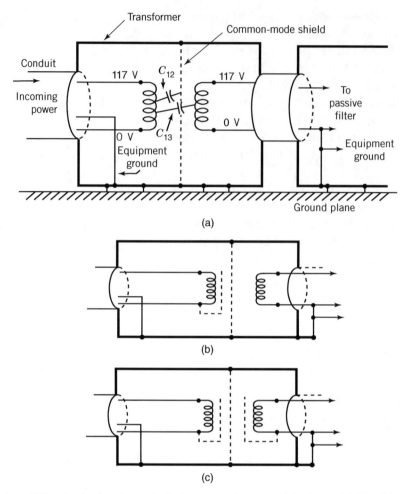

Figure 5.5   A single-phase isolation transformer. (a) One shield, (b) two shields, and (c) three shields

The first or primary shield limits the flow of common-mode noise current in capacitance C_{13}. With a shield, the current now flows in C_{12}. The current that flows in C_{12} flows in coils of the primary and by transformer action this noise can appear on the secondary side. To limit this coupling path, a primary shield is added. This is shown in Figure 5.5.

A third shield is often added on the secondary coil. This limits noise generated on the secondary side from coupling to the primary circuit. It is important to note that these three shields are connected inside the transformer. These short connections are optimum. The idea of carrying these shields to a central *ground*ing point is not correct.

5.29 GROUNDING AND THE PACIFIC INTERTIE

The Pacific Intertie is used to carry utility power between the Pacific Northwest and Southern California. The Intertie carries the power on two dc lines. The technology was first developed in Sweden. Power can travel in either direction but in this intertie it usually flows south. The present power capacity at the source is 3.1 GW, enough power to supply 2–3 million users. The system was proposed to congress in 1961 by President John F. Kennedy and became operational in 1970. It was said that this system would save Southern California $600,000 a day. Shipping energy generated from Columbia River water is a lot cheaper and a lot cleaner than burning fossil fuel. The reason why dc is practical has to do with distance. At 60 Hz, losses related to skin effect and supplying reactive energy make an ac system impractical for distances over 500 miles. With improved inverters, this distance is now much shorter. Worldwide, many system expansions are being planned around these high-voltage dc systems.

The intertie converts three-phase 60 Hz power to plus and minus 500 kV dc. The Pacific Intertie is a two-conductor dc transmission path that goes from the Celilo Converter Station in Washington State to the Sylmar Conversion Center near Los Angeles, a distance of 1362 km. The earth (the voltage midpoint) is used as the neutral or reference conductor. The current in each conductor at capacity is 3100 A. An imbalance of about 10% means an earth current of 300 A. To limit losses in connecting to the earth path, the contact resistance must be in the order of milliohms.

The earth connection at the northern end of the line is in Rice Flats Oregon some 6.6 miles away. The earth connection consists of 1067 cast iron anodes in a 2-mile circumference 2-ft-wide trench filled with petroleum coke. The earth connection is carried to Celilo on two steel-reinforced aluminum conductors 1.1 in. in diameter.

The earth connection in Southern California is equally unusual. A total of 24 silicon iron rods suspended in concrete enclosures are submerged 1 ft above the ocean floor at Will Rogers State Beach. The distance to Sylmar is 30 miles. The two earth conductors are again steel-reinforced aluminum 1.1 in. in diameter. Here are a few facts of interest. The efficiency of the system has been estimated to be at about 77%. The power conductors used 8600 tons of aluminum. The resistance of each dc power line conductor is 0.027 Ω/km.

The converters in use today are banks of light-activated thyristors called valves. These converters are more efficient than the earlier mercury-arc valves. The conversion efficiency that is now available makes it practical to consider shorter runs. In 2010, a 53-mile dc power link was completed between Pittsburg, California, and San Francisco. The line can deliver 400 MW at 200 kV dc or 40% of the peak demand.

5.30 SOLAR WIND

The sun is constantly ejecting charged particles, mostly electrons and protons. The earth's magnetic field deflects most of this current. During periods of intense solar activity the flow of particles can disrupt communications on earth and on rare occasions can cause power outages. The plasma couples dc current flow to long transmission lines that can saturate the cores of distribution transformers. When this happens, the power must be interrupted to avoid system damage. If the system cannot reconfigure from this loss of transmission there will a system blackout that can last for many hours. Shutting down is obviously preferred to replacing hardware. During periods of intense solar activity, the utilities are on constant alert.

Long metal pipelines can couple to this solar wind. In the arctic, long oil pipelines are supported on towers that make connection to the thawed earth below. An insulator must interrupt the conducting loops that are formed or the electrolysis that results from current flow can destroy the pipeline.

CHAPTER 6

Radiation

OVERVIEW

This chapter discusses radiation from circuit boards, transmission lines, conductor loops, and antennas. The frequency spectrum of square waves and pulses is presented. Matching of impedances is required to move energy from a transmission line into an antenna so that it can radiate this energy into free space. Common-mode and normal-mode coupling of fields to conductors is considered. The concept of wave impedance and its relation to shielding is considered. Interference can be analyzed by using a rise-time frequency to represent pulses or step functions.

Effective radiated power from various transmitters is presented. The field intensities for lightning and electrostatic discharge (ESD) are given. Loops generate low-impedance fields that are often difficult to shield. Simple tools for locating sources of radiation are suggested.

See Appendix A for a review of decibels that are used in this chapter.

6.1 HANDLING RADIATION AND SUSCEPTIBILITY

Engineering involves applying scientific principles to the practical world. When electromagnetic radiation and susceptibility are involved, many simplifications are needed to describe what is taking place. It is possible to apply theory to an antenna design, but it is very difficult to apply antenna theory to the radiation from a printed circuit board. It is difficult to calculate the structure of a wave that penetrates a round hole in an infinite conducting plane. If the hole is in a metal box, then a solution might be approximated. If the metal box has internal electronics, then the internal field must be estimated. It is often suggested that the field in the box can be measured using a sensor of

Grounding and Shielding: Circuits and Interference, Sixth Edition. Ralph Morrison.
© 2016 John Wiley & Sons, Inc. Published 2016 by John Wiley & Sons, Inc.

some sort. The difficulty here is that the sensor and its wiring becomes a part of the problem.

N.B.

Almost every practical problem involving radiation or susceptibility requires simplification and approximation.

In the real world, radiation and susceptibility problems are complex. Holes are not round and centered, radiation does not arrive perpendicular to surfaces, and the fields are not plane waves. Computers can be used to work on these problems, but usually many simplifications are required. In fact, engineering economics usually dictates that the full problem not be considered. To understand what happens in practical problems, we can only apply basic principles and seek a worst-case analysis. With this view in mind, we need to discuss some practical tools.

6.2 RADIATION

In the first two chapters, the electric and the magnetic fields were discussed. We saw that both fields were present as energy was transferred into a capacitor, inductor, or through a transformer. In Chapter 3, we saw how waves carried energy on transmission lines. We saw how energy was reflected and transmitted at every transition in characteristic impedance. In this chapter, we will treat electromagnetic fields that leave the confines of a conductor geometry at the speed of light.

In Chapters 1 and 2, we saw that a volume of space has a capacitance and an inductance. For waves traveling in space, electric and magnetic field energy are linked together and share the same space. In a transmission line, the ratio of electric field strength to magnetic field strength had units of ohms. The characteristic impedance of a transmission line turns out to be the square root of the ratio of inductance per unit length to capacitance per unit length.

$$Z_0 = \left(\frac{L}{C}\right)^{1/2}. \tag{6.1}$$

This same ratio applies to electromagnetic energy traveling in space. In this case, the ratio involves the capacitance and inductance per unit volume in free space. This number is $377\,\Omega$.

To show how this radiation process can work, consider a transmission line where the two conductors are spreading apart. This can be approximated by cascading a series of short transmission lines each with an increasing characteristic impedance. Consider a step voltage applied to this line. At every interface, there is a reflection and a transmission. It does not take much spreading of the line before the characteristic impedance approaches 377 Ω. At this point, the reflected wave energy is near zero and the transmitted wave energy is near 100%. It takes many hundreds of reflections and transmissions at all the interfaces to see what happens. After a period of time, a smooth pulse traveling at the speed of light leaves the end of the line and enters space. We have built a transmitting structure. The radiation in this case is a rounded pulse. If the driving voltage is a sine wave, then the radiated field will also be a sine wave.

6.3 SINE WAVES AND TRANSMISSION LINES

In Chapter 3, it was convenient to use step waves to carry energy on circuit traces. Transmission lines are often used to transport sinusoidal wave energy. In AM radio, sine waves are amplitude modulated by an audio signal and carried to an antenna for radiation. In FM radio, the sine waves carried by the transmission lines are frequency modulated. Coaxial cable is a transmission line for transporting carrier-type signals. For the moment, we are interested in the transport and radiation of sinusoidal energy.

Consider a length of transmission line connected to a sinusoidal voltage. We are not interested in transient phenomena only the steady state condition. The voltage waveform along the entire line is a sine wave. If the line is terminated in its characteristic impedance, then there is no reflection and all of the forward energy is absorbed. If the line is open circuited, then what happens depends on the exact line length. We know there is energy stored on the line and that in a steady-state situation, the amount of stored energy is fixed. If the reflected wave appears at the input in phase with the drive voltage, then no input current will flow. This means the input impedance is infinite. For different line lengths, the input impedance will be purely reactive (such as a capacitor or inductor). This means that for a length of unterminated ideal line, the energy that is stored is fixed depending on the voltage level. There is current flow, but energy is not dissipated.

We have already seen that if the transmission line is terminated in a matching impedance (in this case, a matching resistance), the energy moving on the line is dissipated in the resistor. In Chapter 3, we learned that the characteristic impedance of a transmission line is basically a resistance.[1] We are interested in connecting the transmission line to an antenna that can radiate energy into space. To couple energy efficiently into free space, the transmission line and the antenna must look like the impedance of free space or 377 Ω. These radiators (antennas) make our modern society run. Without these radiators, we would not have television or cell phones. If a cell phone is 9 cm long, this is a quarter wavelength at a frequency of 832 MHz. This is approximately the frequency used in a cell phone receiver.

6.4 APPROXIMATIONS FOR PULSES AND SQUARE WAVES

The radiation generated by a circuit may come from a square wave clock or digital logic. Dc-to-dc converters that use square wave circuitry can also radiate. The ability for these signals to cross couple or radiate depends largely on rise time, amplitude as well as on loop area. We start this discussion by examining the frequency content of square waves.

A square wave of voltage or current generates square waves of E and H fields. The sine wave frequencies and their amplitudes that make up a square wave can be derived from a Fourier analysis. These amplitudes are shown in Figure 6.1. The fundamental frequency has an rms amplitude of $2A/\pi$, where A is the height of the square wave. The sine waves that make up the square wave include all the odd harmonics of the fundamental. The amplitude of the third harmonic is one-third the amplitude of the fundamental and the amplitude of the fifth harmonic is one-fifth the fundamental and so forth. These amplitudes and frequencies are plotted on logarithmic scales in Figure 6.2.

Note that in this plot, the harmonic amplitudes lie along a straight line. This line has a slope of 20 dB per decade. We shall refer to this line as an envelope of peak amplitudes. When the square wave has a finite rise and fall time, the Fourier analysis is a bit more complex. The harmonic content is shown in Figure 6.3. In this case, the harmonic amplitudes vary. When the harmonics are plotted on a logarithmic scale, a general form emerges. The harmonic amplitudes are less than a linear envelope out to a frequency $1/\tau_r$, where τ_r is the rise time of the square wave. Above this frequency, the amplitudes are contained by an envelope that falls off as the square of frequency. On a logarithmic

---

[1] There is no dissipation in this resistance. It defines the ratio of voltage to current flow at any frequency for an infinite line.

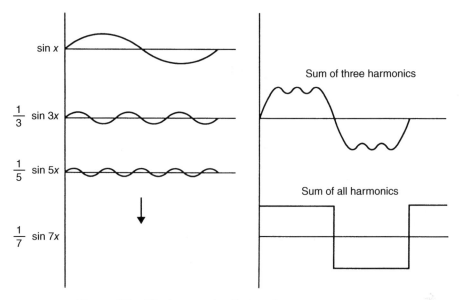

Figure 6.1   The harmonics that make up a square wave

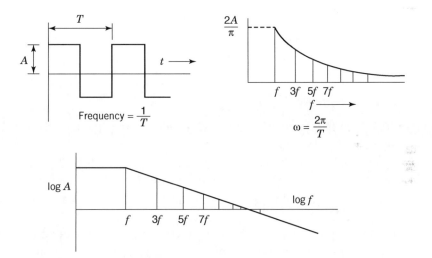

Figure 6.2   The harmonics of a square wave plotted on logarithmic scales

plot, this second envelope has a slope of 40 dB per decade. This plot is shown in Figure 6.4.

For repetitive pulses with a short duty cycle and a finite rise time, a Fourier analysis shows that the harmonics less than the frequency $1/\pi\tau_r$ have an amplitude less than $2A\delta$, where δ is the ratio of pulse time to duty cycle time or $\delta = \tau/T$. A logarithmic plot of the harmonics with an enclosing envelope is shown in Figure 6.5. A worst-case envelope for

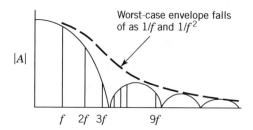

Figure 6.3 The harmonics that make up a square wave with a finite rise time

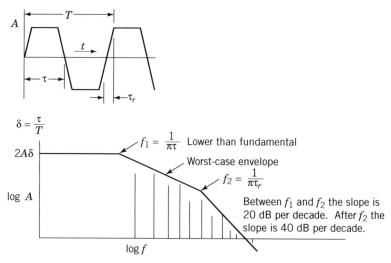

Figure 6.4 The harmonics of a square with finite rise time plotted on logarithmic scales

harmonic amplitudes has a slope of 20 dB per decade from a frequency $1/\tau$ to $1/\tau_r$. Above this frequency, the harmonic amplitudes fall off at 40 dB per decade.

For a single pulse, there is frequency content at all frequencies. The amplitudes are a constant out to a frequency of $1/\pi\tau_r$. This is shown in Figure 6.6.

When a square wave of voltage, current, or field couples to a circuit, the response can be calculated by summing the responses to each harmonic. In most circuits, the coupling process is proportional to frequency. Because the harmonic amplitudes fall off with frequency and the coupling increases proportional to frequency, the two effects cancel. The result is a reconstructed square wave using harmonic content out to a frequency of $1/\pi\tau_r$. When there is a finite rise time, the harmonics above the frequency $1/\pi\tau_r$ are attenuated and usually can be ignored.

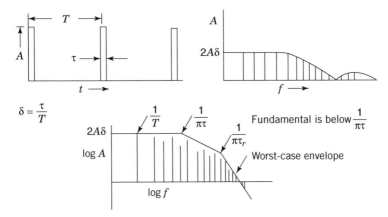

Figure 6.5 The frequency spectrum for repetitive short pulses

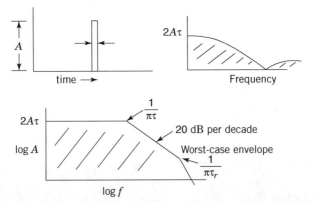

If A is in volts and τ is in microseconds then $2A\tau$ is in volts per megahertz

Figure 6.6 The frequency spectrum of a single pulse with a finite rise time

To analyze how a circuit responds to a complex waveform, a sinusoidal voltage can be selected at a frequency $1/\pi\tau_r$, where τ_r is the rise time. The rms voltage amplitude of the sine wave should be set equal to $2A/\pi$, where A is the peak voltage amplitude of the wave form. This frequency and amplitude can be used in Equation 5.7 to determine if there will be circuit damage or a radiation level out of specification. This all assumes that the exact waveform of the result is not needed. This approach works for either single events or repetitive waveforms.

The two most damaging pulses are lightning and ESD. Arcing at a contact opening can also be considered a pulse. The frequency that characterizes a pulse-like event is again $1/\pi\tau_r$, where τ_r is the rise time.

Later, we will discuss how electromagnetic energy enters an enclosure and how it couples to circuitry. Since the substitute interference is a sine wave and only an approximation, it is wise in design to provide for an additional safety factor from three to five depending on criticality.

N.B.

Lightning pulses have a rise time of approximately 0.5 μs. A peak current level of 100,000 A is the worst case. The sine wave frequency to use is 640 kHz.

N.B.

An ESD pulse has a rise time of about 1 ns. A peak current of 5 A is typical. The sine wave frequency to use is 300 MHz.

N.B.

Pulse-like events often have different rise times and fall times. The shortest time should be used in an analysis.

6.5 RADIATION FROM COMPONENTS

The assumption that is made in circuit theory is that the field energy stored by a component is located in the component. In any circuit analysis, it is further assumed that the stored energy in a component is returned to the circuit. In Chapter 3, we discussed energy supplied to a transmission line. If a transmission line is unterminated, this energy is stored in a distributed capacitance. This energy cannot be returned to the energy source. It must eventually be radiated or turned into heat. The simple explanation is that we are dealing with conductor geometries not idealized circuit components. Here the simplifying rules of circuit theory are not used.

In a parasitic sense, any geometry of conductors is a circuit. Moving energy in this geometry requires moving fields and often these fields are not fully confined. The path a field takes always spreads out the stored potential energy. If the conductor geometry permits, some of this energy radiates. In circuits that process sine wave voltages, radiation can be approximated by considering the time it takes for energy to return to the circuit. This delay is due to the speed of light. The sine voltage that represents the E field can be divided into delayed and nondelayed components. The component of energy that is delayed in time by 90 electrical degrees cannot reenter the circuit and is radiated. If square waves are involved, the luxury of using phase shift to determine a radiated component is not present. A square wave voltage consists of a series of odd harmonic voltages and each harmonic can be considered a separate radiator. The radiated waveform can be constructed from these harmonics. It is not obvious what the radiated pattern will be.

> **N.B.**
>
> Engineers make voltage measurements not energy measurements. Nature moves energy and does not do mathematics. We often have to step back and consider what a measurement means.

6.6 THE DIPOLE ANTENNA

The field pattern in the space around a length of conductor (antenna) can be calculated by assuming a current pattern for the entire conductor. The field at a point in space is then the sum of the contributions from every segment of the antenna. The simplest current pattern to select is a quarter sine wave. Since the tip of the antenna has zero current, we want to pick a frequency where the maximum current flows at the base of the antenna. A good guess says that the antenna length should be one-quarter wavelength long assuming the wave is traveling at the speed of light or $300\,\text{m/}\mu\text{s}$. At $1\,\text{MHz}$, one-quarter wavelength is $75\,\text{m}$. An antenna driven at its base with respect to a ground is called a half dipole. This configuration is shown in Figure 6.7.

The electric and magnetic fields around a dipole are a function of the driving sinusoidal voltage, the directional angle, the distance r from the antenna, and the wavelength. The electric field intensity perpendicular to the antenna can be approximated by

$$E = k_1\left(\frac{\lambda}{2\pi r}\right)^3 + k_2\left(\frac{\lambda}{2\pi r}\right)^2 + k_3\left(\frac{\lambda}{2\pi r}\right) \tag{6.2}$$

where r is the distance to the antenna and λ is the wavelength. The magnetic field can be approximated by

$$H = k_4\left(\frac{\lambda}{2\pi r}\right)^2 + k_5\left(\frac{\lambda}{2\pi r}\right). \tag{6.3}$$

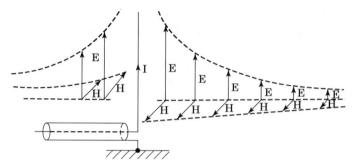

Figure 6.7 A half-dipole antenna

If the antenna is $1/4$ wavelength long, the peak E field near the antenna is $4V/\lambda$.

6.7 WAVE IMPEDANCE

Using Equations 6.1 and 6.2, the ratio of E/H at a remote point is simply k_3/k_5. Since the units of the E field are volts per meter and the units of H field are amperes per meter, the ratio has the units of ohms. The ratio k_3/k_5 is 377 Ω. This does not imply that an ohmmeter can be used to measure the character of space. Since both E and H can be measured over any distance d and the two fields are perpendicular to each other, it is correct to say that E/H equal 377 Ω/sq. This is sometimes written as 377 Ω/□.

At a distance $r > \lambda/2\pi$, the ratio E/H is 377 Ω. At a shorter distance, the ratio E/H will rise proportional to $1/r$. For example, if the frequency is 1 MHz, then $\lambda/2\pi$ is 47.7 m. At half this distance or 23.8 m, the wave impedance is 754 Ω. This impedance value simply means that the E field dominates near this radiating source. The distance $\lambda/2\pi$ is called the near-field/far-field interface distance. This distance will be important when we consider shielding against the penetration of electromagnetic energy. Waves at a distance greater than $\lambda/2\pi$ are called plane waves. Near-fields from loops have a low wave impedance.

The value of Poynting's vector integrated over the surface of a sphere at any distance r from the radiating source must yield the same total power. At half the near-field/far-field interface distance, the E field increases by a factor of $2\sqrt{2}$ and H increases by a factor of $2/\sqrt{2}$. The ratio of E/H is doubled and the power crossing the surrounding sphere is constant.

The radiation from a current loop takes on the same form as Equations 6.1 and 6.2 but with the roles of E and H interchanged. These equations are

$$H = g_1 \left(\frac{\lambda}{2\pi r} \right)^3 + g_2 \left(\frac{\lambda}{2\pi r} \right)^2 + g_3 \left(\frac{\lambda}{2\pi r} \right) \tag{6.4}$$

and

$$E = g_4 \left(\frac{\lambda}{2\pi r} \right)^2 + g_5 \left(\frac{\lambda}{2\pi r} \right). \tag{6.5}$$

For large values of r, the ratio of E/H is again a constant equal to 377 Ω. The near-field/far-field interface distance occurs where $r = \lambda/2\pi$. For values of r less than this value, the wave impedance gets smaller. At one-half the distance, the wave impedance is 188 Ω. At this half distance, the value of E increases by $2/\sqrt{2}$ and the value of H increases by $2\sqrt{2}$. The reduction in wave impedance near the radiating source is the reason it is difficult to shield against the penetration of this field energy. The field near a loop of current where the H field dominates is called an induction field. At power frequencies, the field dominated by current flow is always an induction field. The interface distance at 60 Hz is 500 miles. The wave impedance at a distance of a few inches would calculate to be a few microhms. This number may not be very meaningful, but it does clearly show why shielding this type of field is very difficult.

6.8 FIELD STRENGTH AND ANTENNA GAIN

If a transmitter were to transmit power equally in all directions, the power crossing a surface at a distance r from the radiating source would be

$$P = E \times H \cdot A \tag{6.6}$$

Since the ratio of E to H is 377 Ω, E can be written as

$$E = \frac{(30P)^{1/2}}{r} \tag{6.7}$$

where P is in watts, E is in volts per meter, and r is in meters. For example, the E field 1 km from a 1 MW transmitter is 5.47 V/m.

In most applications, field energy is directed at some target. In radar, the beam is directed outward by a parabolic surface and the beam angle can be a few degrees. In TV broadcasting, the energy is directed at the population and not upward to the sky. Obviously, this directivity greatly reduces the amount of power required to provide field strength at the target. In the case of radar, if the beam has a solid angle of 1°, the power required to produce an E field at a target is reduced by a factor of 360. In the example above, the power requirement would be 2.7 kW. The ratio of 360° to the radiated solid angle is called the antenna gain. The field strength at the target is the same as if a 1 MW transmitter generated a field that propagated uniformly in all directions.

DEFINITION

Effective radiated power. The power level that would provide the required field strength through a solid angle of 360°.

DEFINITION

Antenna gain. The ratio of perceived radiated power to actual radiated power. In the example above, the antenna gain is 360.

In problems involving susceptibility, the peak field strength at the target is all that matters. In a radar signal, the pulse duty cycle may be 1%. This makes it practical to use kilowatts of average power to produce field strengths equivalent to a gigawatt transmitter. The susceptibility problem relates to the field strength as if the transmitter radiated a gigawatt.

The effective radiated power from a list of transmitters is given in Table 6.1.

6.9 RADIATION FROM LOOPS

In this section, we are interested in unintentional radiators. These radiators are the conducting loops that carry signals and power in circuits. For example, when a logic transition sends a signal to a nearby gate, current flows in the loop formed by the voltage source, the logic trace, and the ground plane. If the energy comes from a local decoupling capacitor, there is a current loop formed by the decoupling capacitor, the ground plane, and a circuit trace. Another circuit loop might involve the drive transistors and an associated dc-to-dc converter transformer. A third loop might involve the connection between a triac and a motor.

The electric field pattern from a loop carrying a sinusoidal current depends on many factors including reflections from nearby conductors

Table 6.1 A table of common radiators

Application	Frequency range	Effective radiated power
VLF navigation	10–300 kHz	300 kW
AM radio	0.5–1.5 MHz	50 kW
Fixed HF	3–30 MHz	10 kW
Hams	3–30 MHz	750 W
Land mobile	3–30 MHz	100 W
VHF TV (low)	50–80 MHz	200 kW
FM radio	80–120 MHz	100 kW
VHF TV (high)	150–250 MHz	250 kW
UHF TV	400–900 MHz	5 MW
Radar—military	0.2–100 GHz	10 GW
Radar—ATC		1 GW
Radar—harbor		100 MW

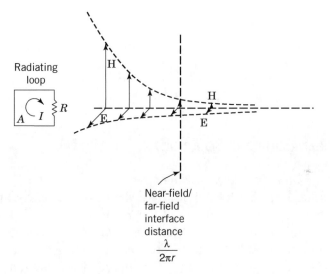

Figure 6.8 The radiated field from a conducting loop

and the angles from the center of the loop. In the spirit of worst-case analysis, we consider the maximum field strength E at a distance r from a radiating loop. See Figure 6.8. Field strength is proportional to loop area, current level, and to the square of frequency. The field strength falls off *linearly* with distance when the field is measured beyond the near-field/far-field interface distance. At 100 MHz, the interface distance is 0.47 m. In equation form, the field strength E beyond the interface distance is approximately

$$E = 6 \times 10^{-3} \left(\frac{IAf^2}{r} \right) \tag{6.8}$$

where E is in decibels microvolt/meter, I is the current in milliamperes, A is the loop area in square centimeters, f is the frequency in megahertz, and r is the distance in meters from the loop. If the distance is 1 m, the loop area is 1 cm², the sinusoidal current is 100 mA, and the frequency is 100 MHz, the radiation is approximately 60 dB μV/m. The loop we are describing may be rectangular, square, or round.

N.B.

Knowing the wave impedance, the magnetic field strength can be calculated from the electric field strength.

If the dimensions of the loop exceed one-half wavelength, then there may be field cancellation. In the spirit of worst-case analysis, the maximum dimension allowed for any one loop is one-half wavelength. This limitation disallows any field cancellation.

6.10 E-FIELD COUPLING TO A LOOP

When a radiated field is associated with any circuit loop, there is a voltage induced in that loop. For analysis purposes, this voltage source can be inserted anywhere in one of the loop conductors. If the voltage is in series with a signal lead, then the interference is amplified as a normal-mode signal. If the interference involves a return conductor sharing several signal conductors, the coupling may be common mode.

The maximum coupling occurs when the direction of field propagation is parallel to the cable direction. In this arrangement, the H field flux of the interfering signal crosses the loop at right angles. The H field converted to a B field can be used to calculate the induced voltage. The rate of change of the magnetic flux yields the voltage from Equation 2.5.

The E field can also be used to calculate the induced voltage in a loop. The E field calculation is simpler because the conversion of units is not involved. At an instant in time, the E field has a different intensity at the ends of the cable. This difference is maximum when the cable length is $^1/_2$ wavelength. The voltages at the cable ends are the E field times the

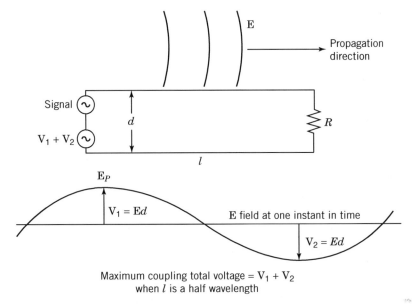

Figure 6.9   Electric field coupling to a pair of conductors

cable spacing d. The maximum induced voltage coupling is twice the peak value of E times d. If the half wavelength is greater than the cable length, the coupling is proportional to the fraction of wavelength. This coupling is shown in Figure 6.9.

6.11 RADIATION FROM PRINTED CIRCUIT BOARDS

The technology used in printed circuit board design and manufacture keeps changing the nature of the radiation problem. In the past, component lead dress was a problem. Today most designs use surface-mounted components, so lead dress is not an issue. There are more ground- and power-plane layers and this has reduced radiation from inner traces. Integrated circuits have gone to ball grid arrays. Large pin counts have made it more difficult to control transmission line fields near the integrated circuit (IC). This has added to the radiation problem leaving the board designer with less margin for error. More vias are being used in designs, but their use has created a new set of problems related to how to transition logic between layers. The error that is made is to provide the current paths and ignore the paths taken by the fields. It must be the other way around. When fields are not provided a controlled characteristic impedance path, the radiation will increase.

   Designers often make the assumption that a ground plane acts as a shield. Unless a conductive surface surrounds a radiating source (no holes), the fields will go in and out of an IC in the spaces between logic leads, power leads, and ground leads. Just adding to the conducting surface is usually not productive.

   The radiation from traces occurs when a wave front is in transition. For steep wave fronts, the radiator can be considered a moving half dipole.

   IC manufacturers must eventually decide whether they will provide energy decoupling inside of their products. So far this action has not taken place.

6.12 THE SNIFFER AND THE ANTENNA

A very practical tool to measure the magnetic fields near apertures or conductors can be built using a section of coaxial cable. The tool called a sniffer is shown in Figure 6.10.

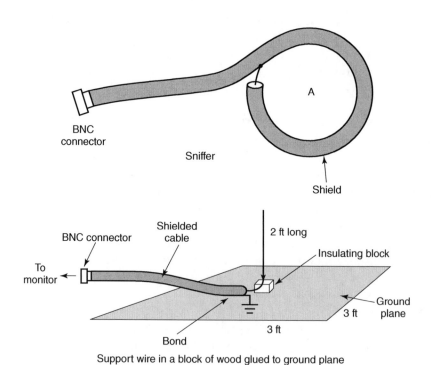

Figure 6.10   A sniffer and a test antenna

The center conductor is connected to the outer conductor to form a shielded loop. The induced voltage is proportional to the loop area A and the rate of change of magnetic flux.

If the frequency is known, the H field can be calculated from the induced voltage. This is not intended as a calibrated tool but as a simple way to locate sources of radiation. Another simple tool that can be built is a small antenna. It too can be used to locate sources of radiation. This structure is also shown in Figure 6.10. The E field can be calculated by noting that the voltage sensed is proportional to the ratio of a half wavelength to antenna half wavelength. The maximum voltage that can be sensed is the E field at a quarter wavelength times the length of the antenna. A smaller version can be built for higher frequencies.

N.B.

Before measurements are made, the ambient field in the area must be considered.

6.13 MICROWAVE OVENS

This kitchen appliance provides a good example of electromagnetic shielding. The space used for cooking is visible through a glass window and accessible through a latched door. The metal mesh in the window serves as a shield to limit radiation into the room. The mesh openings are about 6 mils in diameter. The window screen is bonded to the door frame on the perimeter. These openings form a set of dependent apertures (see Section 7.11). Even the oven light is screened so that radiation cannot leak back out of the oven compartment into the room. The door closes on a pressure gasket or provides a wide metal seam on the perimeter.

The magnetron that generates the electromagnetic field functions a little like an air whistle. An electron beam is passed by an opening in a vacuum cavity that contains a magnetic field. The energy in the oscillating field is coupled into the oven chamber in a waveguide. The frequency of the oscillation is about 2.45 GHz. The wavelength at this frequency is 12.23 cm. This wavelength is effective in coupling the field energy to water molecules in the food. The leakage field is less than the field from a transmitting cell phone.

CHAPTER 7

Shielding from Radiation

OVERVIEW

Cable shields are often made of aluminum foil or tinned copper braid. Drain wires make it practical to connect to the foil. Coaxial cables have a smooth inner surface that allows for the circulation of current and provides control of characteristic impedance. Transfer impedance is a measure of shielding effectivity. Multiple shields, low-noise cable, and conduit each have merits that are discussed.

The penetration of fields into enclosures is considered. This includes independent and dependent apertures, the wave penetration of conducting surfaces, and waveguides. The use of gaskets, honeycombs, and backshell connectors are described. Handling utility power, line filters, and signal lines at a hardware interface are discussed. Methods for limiting field penetration into and out of a screen room are offered.

7.1 CABLES WITH SHIELDS

Decibels are used in many places in this chapter. Appendix A provides a review of this subject.

In analog work, an aluminum foil is often used as a shield around a cable. The foil has a folded seam that runs the length of the cable. The inside of the aluminum foil is anodized to provide protection against corrosion. Because it is difficult to terminate the foil at the cable ends, a drain wire is provided on the outside of the cable foil. This drain wire is made of multistranded tinned copper wires that make contact with the foil along the length of the cable. If the foil should break, the drain

Grounding and Shielding: Circuits and Interference, Sixth Edition. Ralph Morrison.
© 2016 John Wiley & Sons, Inc. Published 2016 by John Wiley & Sons, Inc.

wire connects the segments together. The drain wire is used as the shield connection at the cable ends.

In audio work, where a cable carries a microphone signal a short distance, the cable can be a shielded single conductor. The shield is usually a woven braid, which is more durable than foil. Clip-on or handheld microphones that transmit voice on an rf carrier are rapidly taking over. In instrumentation, best practice requires that the signal common and the shield be separate conductors.

An aluminum foil over a group of conductors provides an excellent electrostatic shield at low frequencies. In analog work, the shield should be connected at one end to the reference conductor preferably where it connects to a ground. If the drain wire is connected to grounded hardware at both ends, then interference can result. Electromagnetic fields in the area will cause current flow in the resulting loop. In a noise-free environment or if the cable run is short, this may not be a problem.

The author has seen cases where one of the cable conductors was dedicated to connecting metal hardware enclosures together. Transient fields from nearby operating relays caused an interference current to flow in this conductor. This coupled field energy to cable conductors, which in turn disrupted logic. This supports the idea that the drain conductor should be kept on the outside of the cable shield. This means the anodized shield surface must be on the inside.

A foil seam does not allow current to flow freely around the cable. Also the foil does not form a very stable geometry. For these reasons, foil shields should not be used where the characteristic impedance of the cable needs to be controlled. The termination of shields at a hardware interface can be critical. A cable terminated by a drain wire allows field energy to penetrate the hardware at the connector. A woven braid can provide a 360° termination depending on the design of the connector.

The term coax is generally applied to cable where the characteristic impedance is controlled. A typical coax cable is a single conductor surrounded by a shield with a controlled geometry. For applications from dc to about 1 MHz, the characteristic impedance of a cable may not be important. Above this frequency, coaxial cable is preferred. The manufacturer supplies specifications relating to signal loss at high frequencies.

The characteristic impedance of a transmission line is a function of the conductor geometry and of the dielectric constant. To transport rf power without reflections, the source impedance, and the terminating impedance must match the line impedance. To transmit high power, the characteristic impedance should be low. In general, high power requires raising the voltage level. Unfortunately, increasing conductor spacing to accommodate a higher voltage is in the direction to raise the characteristic impedance. In many applications, the power level is incidental as the intent is to transfer information at a reasonable voltage level. Obviously, the cable from a transmitter to an antenna should be selected to carry power as well as match impedances. For microwave transmissions, there is no center conductor and high voltages are practical.

Coaxial cable that is used to carry video signals on a carrier must often operate over long distances. Because of distributed reflections, it is undesirable to use a dielectric material in the cable. The coaxial cable used by cable companies has a very smooth inner surface that avoids surface reflections. The center conductor is spaced by a nylon cord that is spiraled around the center conductor. The center copper conductor is an alloy that will not easily kink or bend.

The characteristic impedance of a coaxial cable depends on the ratio of conductor diameters. This is shown in Figure 7.1. This table assumes an air dielectric. If a dielectric is present, the capacitance per unit length is proportional to the relative dielectric constant. Since the characteristic impedance equals $(L/C)^{1/2}$, this impedance depends on the inverse square root of the relative dielectric constant.

The characteristic impedance of open parallel conductors is given in Figure 7.2. As the spacing increases, the characteristic impedance increases. Assume the characteristic impedance of a two parallel conductors spaced by 60 mm is 50 Ω. The characteristic impedance of one conductor spaced 30 mm from a ground plane is simply one-half of this value or 25 Ω. These conductor geometries are shown in Figure 7.2.

In analog work, it is common practice to leave a signal cable unterminated. If the expected bandwidth is greater than a few kilohertz, it is wise to check the frequency response of the cable. Without a termination, there can be peaking in the amplitude response. A series RC termination can shape this frequency–amplitude response. The easiest way to determine the values of R and C is to test the line with a square wave signal at about 3 kHz. An overshoot of about 6% is acceptable. The capacitor should be set to the smallest value that will allow the resistor to control the overshoot.

7.2 LOW-NOISE CABLES

Special low-noise cable is needed in applications that use piezoelectric transducers. When signal cables are flexed, charges are generated when the conductors rub against the surfaces of the dielectrics. This is known as a triboelectric effect. To a charge amplifier, these moving charges are

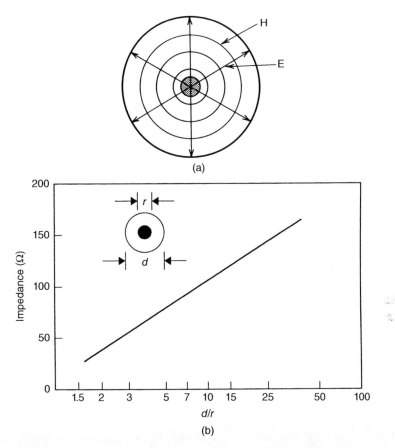

Figure 7.1   (a, b) Characteristic impedance of a coaxial geometry

noise. To limit this effect, conductive materials are added around the dielectric. These low-noise cables are available from the manufacturers of piezoelectric transducers.

7.3 TRANSFER IMPEDANCE

Cables often couple to external electromagnetic field energy. As we have said before, current will take this route because the associated fields store less energy. The coupled energy flows between the cable shield and other nearby parallel conductors. If some of the surface current flow can find a way to the inside surface of the shield, there will be field on the inside of the cable. If there is field there is voltage. This mechanism is called transfer impedance.

L/l	Z (Ω)	H/h	Z (Ω)
1.1	53	0.6	37
1.5	115	1.0	79
2.0	158	2.0	124
2.5	188	2.5	138
3.0	212	3.0	149
4.0	248	4.0	166
5.0	275	5.0	180
10.0	359	10.0	221
30.0	491	30.0	287
100.0	636	100.0	359

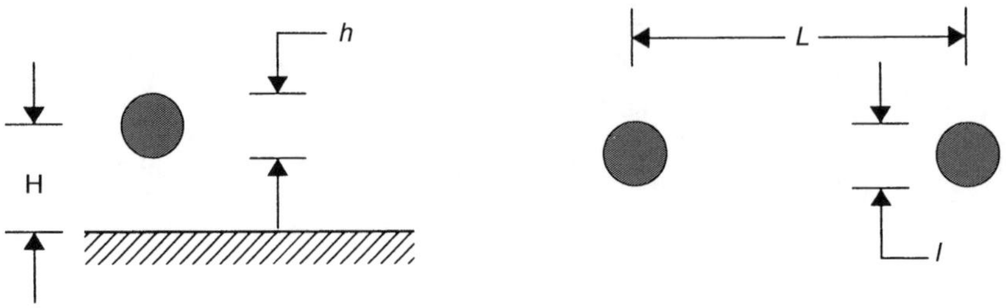

Figure 7.2 The characteristic impedance of parallel conductors

---

N.B.

If an interfering current flows on the inside surface of a coaxial shield, there is an unwanted field inside the coaxial cable that has already coupled to the signal.

The field that couples into a cable through the shield carries energy that moves in both directions inside the cable. If the cable is terminated at both ends, then one-half of the coupled energy is dissipated in each termination. If a current I on the shield causes an interfering voltage V at each termination, the ratio of $2V/I$ is said to be the transfer impedance of the cable. This value is normalized for 1 m of cable. The test for transfer impedance is shown in Figure 7.3.

At low frequencies, the current in a cable shield uses the entire cross-sectional area of the shield. The IR voltage drop in the shield produces an internal field. For solid conductors at frequencies above 10 kHz, the currents tend to stay on the outside surface so there is little internal field. For braided cables, for frequencies above a few megahertz, the current that gets to the inside surface increases with

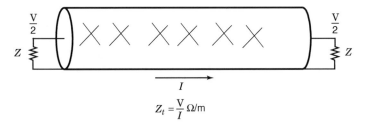

$$Z_t = \frac{V}{I}\ \Omega/m$$

Figure 7.3   Transfer impedance test for a coaxial cable

frequency. The transfer impedance is a function of both the tightness of the weave and the fineness of the braid.

The undulation of the braid adds inductance to the forward path. The current flow in each conductor strand makes many contacts. On average, conductors heading to the outside surface make contact with conductors that are directed toward the inside. As a result, some of the external surface current transfers to the inner surface. If there are two insulated braids, then the coupled energy is reduced. There is field transferred into the space between the braids. This field is then transferred to the center of the cable by a second transfer process. The lowest transfer impedance is achieved by using a solid conductor for a shield. Mechanical flexibility can be provided if the shield is corrugated.

Transfer impedance has units of dB ohms-per-meter. A 0-dB figure means $1\,\Omega/m$ while a 20-dB figure implies $10\ \Omega/m$. The transfer impedances for a few standard cables are shown in Figure 7.4. It is interesting to note that the transfer impedance for some braided cable is high enough to make the shield ineffective above a few $100\,MHz$. Obviously, the amount of coupling increases with cable length. For cables longer than a half wavelength, the coupling tends to cancel. In a worst-case analysis, this cancelation cannot be accepted. The maximum cable length used in any calculation should be one-half wavelength.

In applications where the cable runs are short, the cable type may not be important. As we shall see later, the treatment of the cable at the connectors is often far more critical than the type of cable that is selected.

N.B.

The transfer impedance of some braided cable in the range $300\,MHz$ can be quite high.

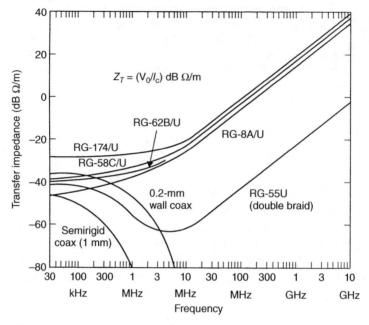

Figure 7.4   The transfer impedance for a few standard cables

7.4 WAVEGUIDES

Electromagnetic fields can propagate in free space at all frequencies. For frequencies below 1 MHz, the antennas that are used to radiate energy into space get very large. When two conductors are present, it is easy to move field energy from point to point in the frequency range dc to perhaps 100 MHz. Waveguides are the preferred method of transporting energy above this frequency. A waveguide is a hollow cylindrical conductor. Consider field energy where the half wavelength is the diameter of the waveguide opening. This wave can establish a pattern of reflections on the inside surface of the guide that will allow it to propagate down the guide. At specific higher frequencies, the waveguide can support the flow of field energy in various field patterns or modes. As the frequency increases, the number of permitted modes increases until there is no longer a modal restriction. Waveguide propagation requires no center conductor. If an insulated conductor is added, the waveguide becomes a poor piece of coax.

The guide attenuates an electromagnetic wave that has a half wavelength that is greater than the waveguide opening. Under this condition, the waveguide is said to be operating beyond cutoff. The attenuation of this type of wave is given by

$$A_{\mathrm{WG}} = 30 \left(\frac{h}{d} \right) \tag{7.1}$$

where the attenuation factor A_{WG} is in decibels.

If this ratio h/d is 3, the attenuation factor is 90 dB. A waveguide beyond cutoff provides a significant attenuation factor. Waveguide attenuation can be used to shield an enclosure at its seams.

FM radio operates at around 100 MHz. The half wavelength is about 1.5 m. Wave energy from an FM station can easily propagate into tunnels and into underground parking structures. An AM radio station might broadcast at 1 MHz. At this frequency, the half wavelength is 150 m and this energy will not propagate into these structures. If an insulated conductor is added at the roof of a tunnel, it becomes a section of coaxial cable. This one conductor will allow the fields from AM stations to enter the tunnel. This one added conductor will also allow signals to radiate from an enclosed structure.

N.B.

One insulated conductor entering a shielded enclosure can allow radiation to enter or exit that enclosure.

7.5 ELECTROMAGNETIC FIELDS OVER A GROUND PLANE

A simple picture of field energy propagating along an infinite conducting surface involves a plane wave. Ideally, if the conductor has zero resistivity, the surface current that flows does not dissipate energy. The only restriction is that the E-field vector must be perpendicular to the surface. The H field must then be directed along the conducting surface, which requires a surface current. Any horizontal E-field component would require infinite surface current.

If an electromagnetic wave arrives perpendicularly to a perfect conducting surface, the wave simply reflects. The reversal of the E field at the conducting surface is similar to the reflection at the end of a shorted transmission line. The reflected wave cancels the voltage at the surface but leaves the current unchanged. For a plane wave reflecting off of a perfect conducting surface, there must be current flow to support the H field at the surface.

For conductive metallic surfaces more than a few millimeters thick, plane waves are reflected and essentially do not penetrate the surface. Plane wave energy will usually enter an enclosure through an aperture involving a hole or a seam, not through the enclosure walls.

N.B.

For conducting surfaces more than a few millimeters thick, very little plane wave energy can propagate through the conductor.

N.B.

Skin depth limits current penetration for high-frequency current flow. Thickness is provided so that the surface is robust enough to be practical.

N.B.

The only way to consider low impedances for large distances is to spread the current flow out over a large conducting surface area or volume.

N.B.

A lightning pulse of 100,000 A flowing evenly over a sheet of copper 1 mm thick covering an area of 1000 ft^2 would cause a voltage drop of about 1.5 V. This same pulse flowing in 100 inches. of #00 wire would result in a voltage drop of over 5,000,000 V.

7.6 FIELDS AND CONDUCTORS

The current that flows in a conductor flows because there is an E field in the conductor. The E field accelerates the electrons, but they give up their energy when they collide with the atoms in the conductor. As a result, the electrons reach an average velocity, which we interpret as current. Consider a #19 copper conductor, which has a resistance of 8 Ω/1000 ft. In 1 m, the resistance is 26.2 mΩ. To support a current flow of 1 A, the voltage drop is 26.2 mV. This is an E field of 26.2 mV/m. In a 10-V circuit where the spacing is 1 cm, the E field between the conductors is 1000 V/m. The ratio of tangential E field to the perpendicular E field is approximately 38,000 to 1. This supports the idea that there is very little tangential E field in most of the circuits we consider.

Consider a large conducting plane where current enters and leaves at two points. The concentration of current at the points of contact depends on both the area of the connection and skin depth. Because the current concentrates at the points of contact, the ohms-per-square concept no longer applies.

At high frequencies, a significant electromagnetic field can exist near the points of contact. The H field dominates at the contact point and therefore it appears as a series inductance.

7.7 CONDUCTIVE ENCLOSURES – INTRODUCTION

In this section, we will discuss how electromagnetic energy can enter or leave a conductive enclosure. The energy can couple through and around cable connections, through apertures, and through the conducting enclosure itself. For a near-induction field, a conducting enclosure may be an ineffective shield allowing direct field penetration.

Earlier we discussed filtering an input circuit to limit interference coupling. In some analog circuits, the effects of radiation are subtle. Out-of-band fields can cause loss of loop gain and add offsets. If local filtering is effective, then enclosure shielding may be unnecessary. This chapter discusses field mechanisms without regard to application.

The radiation that enters an enclosure will couple to internal electronics. Before we discuss enclosure shielding, a story might help to set the stage.

STORY

Consider a metal tub with a lid. Place a small battery-operated FM radio in the tub and note that it receives signal. Now place the lid on the tub. Place your ear near the lid and listen for signal. The radio should stop playing. Now place a section of insulated wire into the tub with a few feet of wire dangling on the outside. The radio should again receive signal.

N.B.

It only takes one unfiltered conductor to violate a shield enclosure.

The rule is simple. Do not attempt to shield an enclosure unless you intend to block all points of field penetration.

When the source of interference is unknown, proceed to experiment by leaving all fixes in place. When the problem is solved, then remove the fixes one at a time. It can happen that there are two or more entry points. If the fixes are tried and then removed, the solution may not be found.

N.B.

A boat with many holes will sink if there is one unplugged hole.

7.8 COUPLING THROUGH ENCLOSURE WALLS BY AN INDUCTION FIELD

Consider a copper or aluminum enclosure. For power-related induction fields (60 Hz and harmonics), there is little reflection loss and the field enters unattenuated. There can be attenuation when the enclosure is made from a material having a high permeability. The best magnetic material is Mumetal®. The high permeability is obtained by annealing a magnetic alloy in a magnetic field in an inert atmosphere. The physical size of a Mumetal part is limited by the practical size of magnets that generate fields in the annealing oven. After a part is annealed, it cannot be punched, drilled, or bent, or it loses its permeability.

Shielding against induction fields can be accomplished by using layers of steel and copper. For example, signal transformers can be housed in nested cans of iron and copper.

The permeability of some magnetic materials can be as high as 100,000. This permeability is measured at a maximum flux density for the material. For this same material, the permeability may drop to 1000 at 100 G and at the milligauss levels the permeability may be as low as 2 or 3. This low permeability shows how difficult it is to shield against low-intensity power frequency induction fields.

Magnetic fields can be reshaped by the presence of magnetic materials. In some situations, this reshaping can reduce magnetic coupling to

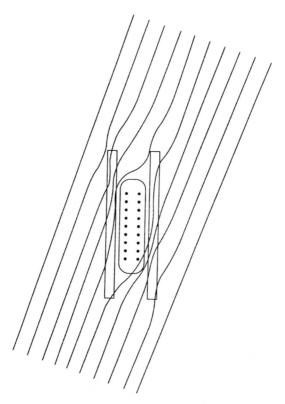

Figure 7.5   Diverting a magnetic field

an acceptable level. An example is shown in Figure 7.5. Do not expect a significant attenuation factor with this approach.

If a cable is coated with a layer of permeable material, an external field can be diverted so that the magnetic flux does not cross the area between two signal conductors. The attenuation factor is usually less than 20 dB depending on the field strength and the permeability of the magnetic material. Twisting the signal pair can also be used to reduce this type of normal-mode coupling. The coupling for a half twist cancels the field coupled in the next half twist. Obviously, twisting is not available if the cable is coax.

7.9 REFLECTION AND ABSORPTION OF FIELD ENERGY AT A CONDUCTING SURFACE

When an electromagnetic field impinges on a conducting surface, two mechanisms occur. Part of the arriving energy is reflected. The fraction that is not reflected enters the conductor and is attenuated by skin effect. For conductors that are more than a few millimeters thick, this field energy is simply converted to heat. For thin conducting layers, some of this energy may actually penetrate the conductive barrier. The reflection loss can be approximated by Equation 7.2, where Z_W is the wave impedance

$$R_{dB} = 20 \log \left(\frac{Z_W}{4Z_B} \right) \tag{7.2}$$

and Z_B is the barrier impedance. Note that for near-induction fields, Z_W is very low and there is essentially no reflection. The reflection loss can never become negative as this would imply a reflection field greater than the arriving field.

Either the arriving E or H field can characterize the wave energy that enters an enclosure through the conductive walls. If the reflection loss is 40 dB and the thickness is 2 skin depths, the arriving field is attenuated by a total of 57.4 dB. If the arriving E field is 40 dB V/m, the E field inside the enclosure is 40–57.4 dB or −17.4 dB V/m. In the spirit of a worst-case analysis, this field strength is not a function of enclosure volume, field direction, or polarization. In equation form

$$F_{INT}(dB) = F_{EXT}(dB) - R(dB) - 8.68n(dB) \tag{7.3}$$

where F_{INT} is the internal field, F_{EXT} is the external field, R is the reflection loss, and n is the number of skin depths of the material. For pulses or square waves, the skin depth is calculated based on the frequency $1/\pi\tau_R$, where τ_R is the rise time.

Fields generated inside of an enclosure can radiate out through the enclosure walls. If the radiator has a low wave impedance, there may be little reflection loss. In this situation, the enclosure may not be effective in attenuating field energy. Equation 7.3 applies for both directions of wave transport.

7.10 INDEPENDENT APERTURES

Field energy can enter an enclosure through an aperture in any conductive wall. The wave pattern on the inside of the enclosure depends on many factors including the presence of internal hardware, the shape of the aperture, the polarization and direction of the field, and the size of the enclosure. A worst-case analysis assumes that the field intensity is not attenuated if the aperture is greater than a half wavelength in dimension. It is common practice to assume that the wave impedance inside the enclosure is the same as the arriving field.

For apertures greater than a half wavelength in dimension, it is best to assume no attenuation. For dimensions smaller than a half wavelength, the field attenuation is assumed to be the ratio of half wavelength to aperture opening. As an example, if the half wavelength is 20 cm and the aperture opening is 2 cm, the attenuation factor is 20 dB or a factor of 10.

A single aperture might be a seam. The dimension to consider is the seam length. Even if the seam appears to be optically tight, it is still an aperture. The reason for this assumption relates to surface current flow. If the seam interrupts the flow of surface current, it acts as an aperture. Closing a seam requires the use of a conducting gasket. See Section 7.17.

Multiple apertures that allow the free circulation of surface current around each aperture will allow multiple points of field entry. The field strength inside the enclosure is assumed to be the sum of individual penetrations. If an external field is 10 V/m and one aperture attenuates this field by 40 dB and another aperture attenuates the field by 46 dB, the two field strengths are 0.1 and 0.05 V/m. The sum of these fields is 0.15 V/m. Note that the decibel measure of field strength is not additive. A field strength of 0.15 V/m is −16.5 dB V/m. The initial field is 20 dB V/m. The attenuation factor expressed in decibels is 36.5 dB.

DEFINITION

Independent aperture. An opening in an enclosure that allows the free flow of surface current around the enclosure.

N.B.

When multiple independent apertures allow field entry, the internal field can never be greater than the external field.

7.11 DEPENDENT APERTURES

Arrangements of apertures that do not allow the free flow of surface current are called dependent apertures. An example might be a group of closely spaced ventilation holes. The radiation that penetrates a group of dependent apertures is the same as if there is one aperture.

A wire mesh or screen is considered a set of dependent apertures as current cannot flow freely around each opening. The aperture size for a screen is the dimension of one opening. There are two restrictions: (1) The grid of conductors that make up the grid must be bonded at each crossing. For example, screening made from aluminum conductors can oxidize. When this happens, the aperture openings are not controlled. (2) The screen or mesh must be bonded to a conducting surface along its entire perimeter. If there is no bond, the diameter of the entry way becomes the aperture.

Consider an enclosure made from conducting panels that are held together by screws. The seam between two screws is considered an aperture. Because surface currents cannot flow freely around each screw, the apertures are considered dependent. For this reason, the seams appear as one aperture where the maximum spacing between screws is the dimension of the aperture.

If a 20 dB improvement in field penetration were required, the number of screws would have to be increased by a factor of 10. The practical solution to this problem is to close the apertures using a conducting gasket. To be effective, a gasket material should make a continuous connection along the seam. Gasket contact areas should be plated so that there is no chance of oxidation.

If the enclosure is made of sheet metal panels, the edges can be bent to form a flange. This flange can be used as a waveguide opening. If the flange makes many wide surface contacts, the aperture openings are small but they are deep. This type of aperture is considered a waveguide beyond cutoff. Even if the openings are independent, the attenuation can be significant. See Equation 7.1.

If panels are made of molded plastic with conductive surfaces, then flanges can be a part of the design. These flanges can be used with screws or gaskets to form waveguides beyond cutoff. If ventilation holes are plated and extended in depth, then waveguide attenuation is available.

7.12 HONEYCOMBS

Honeycomb structures are often used to ventilate an area and provide for field attenuation. The honeycomb is formed from conducting hexagonal cylinders that are bonded (flow soldered) together. Consider an impinging field at 100 MHz with an E-field intensity of 20 V/m. A half wavelength at this frequency is 1.5 m. The field at each opening is attenuated by the ratio of the opening to the half wavelength. If the openings are 1.5 cm, the field strength at each opening is 0.2 V/m or −14 dB V/m.

A wave propagates down each honeycomb cell. If the cells are 4.5 cm long, the attenuation in each cell using Equation 7.1 is 90 dB. The field entering the enclosure from each cell is thus −104 dB V/m. The cells in a honeycomb are independent apertures as current can circulate freely inside each cell. The field that penetrates the enclosure is the sum of the fields from each cell. If there are 20 cells, the internal field strength is increased by 26 dB to −74 dB V/m. Converting from decibels, the internal field is 0.0002 V/m.

N.B.

If the perimeter of the honeycomb is not bonded correctly to the mounting surface, then wave energy can enter through any resulting aperture. In the above example, a 1.5-cm opening would allow the internal field strength to be 0.2 V/m. The honeycomb would be ineffective.

Most honeycomb filters are supplied with mounting hardware and a gasket. The mounting surface should be plated to avoid oxidation. The surface must not be painted or anodized. If the honeycomb is removed for cleaning, the gasket may need replacement if there is any question about a good contact over the entire surface.

7.13 SUMMING FIELD PENETRATIONS

The fields that enter an enclosure can enter through apertures, directly through the skin or directly on conductors. These conductors might be used for input, output, shielding, control, or power. These same conductors can carry fields out of an enclosure.

N.B.

Nature does not read labels or color codes. Energy can flow in both directions on every conductor entering an enclosure.

A conductor that enters an enclosure with a voltage V creates an E field that is determined by conductor spacing. Conductors that carry power can direct field energy originating from other hardware or the environment into the hardware. These same conductors can carry field energy to other pieces of hardware. For this reason, it is important to consider every conductor that enters or leaves an enclosure. If the conductor is carrying current, the H-field intensity and Faraday's law can be used to determine coupling to nearby loops. If the E field is known, then capacitive coupling can be used to calculate coupling.

Up to this point, we have considered fields that penetrate through the skin or through apertures from one interfering source. In a worst-case analysis, these fields are simply added together. If fields enter on conductors, they too must be included. If fields have different spectrums, they must be treated as completely separate problems. The fields from different sources can be added together using their rms measure. The total field intensity is the square root of the sum of the field intensities squared.

7.14 POWER LINE FILTERS

Power line filters are used to restrict the flow of interference into and between pieces of hardware. Filters consist of series inductors in the *grounded* and *ungrounded* power conductors and shunting capacitors. Capacitors can go between power conductors or from power conductors to *equipment ground*. The National Electrical Code (NEC) prohibits filter components from being placed in series with the *equipment ground* as this may limit the flow of fault current. Filters are often purchased as readymade components that are mounted on the hardware by the user. It is common practice to manufacture a filter assembly that includes an "on" switch, a power breaker, a power cord connector, and even an "on" light.

Power filters have published specifications that show the attenuation of voltages on the power line. This attenuation data should include both common-mode and normal-mode interference. Normal-mode filtering attenuates voltages between the power conductors and common-mode filtering attenuates voltages between power conductors and *equipment ground*. If the unfiltered power conductors are brought into the enclosure to attach to the filter, these conductors can radiate directly into the hardware and the filter is bypassed. The correct geometry for mounting a filter is shown in Figure 7.6.

Electrical interference can be carried by any conductor pair and one of the conductors can be *equipment ground*. This means that unfiltered power conductors and the *equipment ground* conductor must be kept out of the enclosure. This safety conductor should terminate on the inside of the filter enclosure without electrically entering the hardware

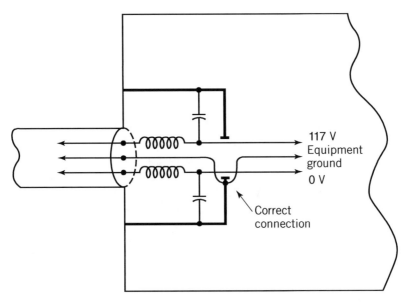

Figure 7.6   The location of power conductors and a line filter

enclosure. If filter currents must flow from inside the filter enclosure to the outside of the filter enclosure to get to *equipment ground*, there must be fields outside of the filter enclosure and this violates the intent of the filter.

As a side note, it is important to realize that at high frequencies, current cannot flow through a piece of metal. Because of skin effect, currents must find a path on conducting surfaces. Consider a filter capacitor that terminates on the inside of a metal can. If the current must flow to an external connection, it must flow through a hole in the can to get to the outside surface. If this current flows on the outside surface of the can, there is an external field. This is what the filter is supposed to eliminate. The presence of this field implies that an impedance has been placed in series with the capacitor. In other words, the filter is compromised.

There are many factors to consider in filter design. The field pattern generated inside a filter enclosure must not couple across the filter circuit. This may require some internal partitioning. All inductors have a resonant frequency. Above this frequency, the inductor looks like a capacitor. In this frequency range, the filter looks like a capacitive divider. All shunt capacitors have a series inductance, which is also in the direction to limit filter performance.

There are many filter types. The filter can be L shaped, pi shaped, T shaped, or a combination of all the three. Filters can be applied to one or both power conductors. Simple line filters are L shaped with the inductors facing the source of power. The line-to-line capacitor on the load side provides a low-impedance source for step demands in energy. The inductors on the power side of the filter limit current flow for pulses on the line. The L sections should be reversed if the interference comes from the load side of the filter.

Power line filtering starts at frequencies generally above 100 kHz. This limits the size of the filter elements and provides filtering at frequencies where there might be radiation. Line filters are not intended to limit harmonics of the power source. Filters that perform this function are very large and expensive.

N.B.

Filters that attenuate signals over a wide frequency range are usually built in sections. This obviously adds to the cost.

If the filter uses a plastic housing or if the filter snaps into an opening using spring clips, the *equipment ground* connection to the hardware must be made separately. If the *equipment grounding* jumper is brought inside the hardware enclosure, it should be kept short to limit loop area and radiation. This method of mounting a filter is not recommended.

For best performance, filter housings should bond to the enclosure. Painted or anodized surfaces should not be used. To avoid oxidation, these surfaces should be plated. In some applications, it may be necessary to use a conducting gasket between the filter housing and the enclosure.

7.15 BACKSHELL CONNECTORS

Conductors that enter an enclosure through a connector can carry interference. This is no different than the problem of interference entering from the power line. Connectors are available with internal filters. The performance of filters in a connector is limited because of the available space.

If an arriving cable is shielded, most of the interference current is carried on the outside surface of this shield. Field energy will enter the hardware through the connector if the shield has any openings at the connector. If the braid is bunched to form individual conductors and

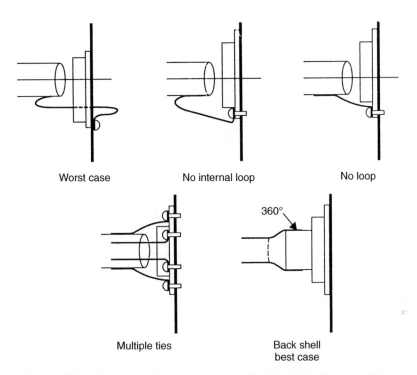

Figure 7.7 Methods of terminating the braided shield on a cable

then terminated on mounting screws, interference current will flow on the surfaces of these formed conductors. This allows field to couple to the cable conductors and enter into the enclosure. Ideally, shield currents should stay on the outside surface of the shield and flow in a smooth manner to the outside surface of the enclosure. This is the purpose of a backshell connector. It terminates the shield through 360° at the connector. The connector is then mounted using a gasket to close any possible aperture. The termination of a shielded cable using a backshell is shown in Figure 7.7. In a noisy environment, this is the only way to keep field energy from entering an enclosure at the connector.

N.B.

Bayonet Neill–Concelman (BNC) and threaded Neill–Concelman (TNC) connectors terminate the coax shield so that current flows uniformly to the terminating enclosure surface.

7.16 H-FIELD COUPLING

The fields that enter an enclosure can couple into a circuit through conductive loops. The largest loops often involve cables between pieces of hardware. If the E field is known, the H field can be determined from the wave impedance. Once the H field is known, the B field can be calculated using the relationship

$$B = \mu H \tag{7.4}$$

where μ is the permeability of free space or $4\pi\,10^{-7}$. The B field flux ϕ is the B field intensity times the loop area in meters squared. The voltage is given by

$$V = \frac{d\phi}{dt}. \tag{7.5}$$

If the original E field is given as an rms value, the voltage V will also be an rms value.

Example 7.1.   Consider an E field of $10\,\text{V/m}$ at $100\,\text{MHz}$ inside an enclosure. The H field assuming a plane wave is $10/377 = 0.027\,\text{A/m}$. The B field intensity is found from Equation 7.4 or $B = 3.33 \times 10^{-8}\,\text{T}$. If the coupling loop is $0.01\,\text{m}^2$, then the flux is 3.33×10^{-10} webers. The induced voltage is $2\pi f$ times this flux or $0.2\,\text{V}$.

Example 7.2.   An electrostatic discharge (ESD) pulse strikes an enclosure $10\,\text{cm}$ from an aperture $10\,\text{cm}$ long. What is the voltage induced in a loop $10\,\text{cm}^2$ near this aperture? Assume a 5-A pulse where the rise time implies a frequency of $300\,\text{MHz}$. Since $2\pi r H = 5\text{A}$, H is equal to $7.96\,\text{A/m}$. B is equal to $10^{-5}\,\text{T}$. At $300\,\text{MHz}$, the half wavelength is $0.5\,\text{m}$. The wave is attenuated at the aperture by a factor of 5. The field intensity B inside the enclosure is $0.2 \times 10^{-5}\,\text{T}$. The flux in webers is BA or 0.2×10^{-8} webers. The voltage induced at $300\,\text{MHz}$ is $3.7\,\text{V}$. If this voltage adds to a logic level, it could damage an integrated circuit.

7.17 GASKETS

Conductive gaskets are used to close apertures. One construction method is to embed filamentary lengths of stainless steel in a carrier. When the gasket is under pressure, it makes many connections to the conductors involved to close the aperture. Gaskets are available as patterned parts or as strips of various widths. Some gaskets are in the form of a metal-braided rope with sharp edges. This type of gasket can be placed in a preformed groove to connect two pieces of metal. When the metal parts are assembled, the gasket closes the aperture.

N.B.

Surface preparation is key to the proper installation of a gasket.

N.B.

Gaskets often deform when installed. Gaskets that are removed when equipment is serviced may have to be replaced.

N.B.

The shielding effectivity of gasket material is supplied by the manufacturer. Shielding effectivity is the ratio of field penetration before and after the gasket material is in place. The manufacturer should supply information on how the test was performed.

Metal cloth is available as a shielding material. To close an aperture, the cloth must be bonded around its perimeter. Since cloth is somewhat fragile, it should not be used where it might be damaged.

7.18 FINGER STOCK

A form of gasket involves finger stock. It is often necessary to shield the aperture around a door. The seam length on a door is long, so it is necessary to make many contacts between the door and the doorjamb. Finger stock as the name suggests provides many contacts in parallel. If the individual connections are deep, the openings take on the character of waveguides. Typically, beryllium copper fingers make contacts that are about a quarter of an inch wide with eighth-inch separation. The stock should be mounted inside a folded cover so that the fingers do not catch on clothing.

7.19 GLASS APERTURES

A thin conductive layer can be plated onto glass to attenuate electromagnetic fields. Unfortunately, these same conductive materials also attenuate light. For this reason, any solution is usually a compromise. Contact must be made to the conductor around the perimeter of the glass so that this glass aperture can be closed.

N.B.

If the optical path includes a waveguide beyond cutoff, the radiation can be effectively controlled in either direction.

If the opening is 10 in. wide, a 10-in.-long hood bonded to the opening can provide 30 dB of field attenuation out to 1 GHz.

7.20 GUARDING LARGE TRANSISTORS

The collectors or drains for power devices are often connected to the metal housing. When these devices are used to handle large amount of power, it is necessary to provide for the flow of heat. One method is to mount these transistors on a large conducting surface. To avoid an electrical connection, a thin insulating gasket is added between the transistor and this surface. To increase the heat conducting area, a thin layer of thermal conducting paste can be applied to the gasket.

The capacitance from the transistor housing to the conducting surface can allow current to circulate in the *equipment ground* system. To limit this current, the mounting gasket can be metal with insulation on both sides. This metal forms a guard shield that can be connected to circuit common. This shunts most of the parasitic current back to the circuit rather than out into the facility. A typical circuit is shown in Figure 7.8. The leakage capacitance around the gasket might be 5 pF while the capacitance without the guard might be 50 pF. If the circuit is arranged so that the drains or collectors are at ground potential, this guarding is not required.

7.21 MOUNTING COMPONENTS ON SURFACES

At high frequencies, skin effect keeps currents on the surface of conductors. When components are mounted on conducting surfaces, the way

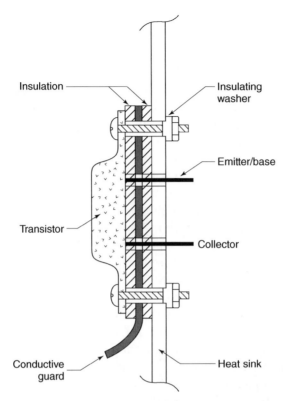

Figure 7.8   A guard gasket applied to a transistor

conducting surfaces mate is important. If the current concentrates at several points, the result is an inductive connection. At low frequencies, this inductance is of no concern. Above a few megahertz, this inductance can influence the performance of the component by introducing feedback or cross coupling. The important point to consider is that there should not be a concentration of current flow if there is a choice. As an example, a component that is riveted in place may function differently than a component that is soldered into position.

The current that flows on the outside of a metal component housing can only enter the housing through a hole in the enclosure. This is another case where knowing the current path and the field pattern is important.

Circuit paths that require sharp changes in direction are inductive. In some cases, the roughness of a surface can add inductance through skin effect. If manufacturers provide recommendations on surface and contact quality, these recommendations should be followed.

7.22 ZAPPERS

A zapper is a testing device that generates high-voltage pulses. These pulses can be used to test a piece of hardware to see if it is susceptible to radiation. Specifically, the test shows whether the hardware is susceptible to ESD. There are two modes of operation. In the first mode of operation, the zapper probe is in contact with the hardware and a pulse of current flows from the probe tip to the hardware. In the second mode, the probe tip is placed near the hardware and the zapper generates an arc to a conducting surface. The pulse rate and the intensity of the pulse are both variables. In the contact mode, the H field dominates near the point of injection. In the arcing mode, the E field is dominant.

There is a close relationship between radiation and susceptibility as both are related to loop areas in the circuitry. In general, if a device is not susceptible to radiation, it probably will not radiate. The test proceeds by using the probe on points such as cables, connectors, seams, displays, and controls. The test should proceed in 1000-V steps from 1000 to 15,000 V. The midrange around 7000 V is important as this is where the intensity of the field energy is apt to peak. Above this voltage, energy is lost in heat. If at any time the unit under test malfunctions, it is wise to stop the test and consider some sort of modification.

If the device under test is insulated or isolated in any way, a discharge path should be provided. This should be done so that there is no accumulation of charge from repeated pulses. A resistor of 100 MΩ can provide an adequate discharge path. Without this path, there can be damage to the circuit that is not related to a susceptibility test. As an example, hardware that is powered from a transformer has no discharge path to earth unless one is provided. Battery-operated hardware also needs a discharge path or the hardware will assume the potential of the zapper.

7.23 SHIELDED AND SCREEN ROOMS

The terms screen and shield are often used interchangeably. A screen can reflect rf energy, but it has no effect on a magnetic field. To reflect near magnetic fields, a thick layer of steel is required. Steel can be both a structure and the shield and when needed implies a more permanent fixture. A room made of steel can provide a space that is electrically quiet so that experiments, processes, or measurements can be carried out. These rooms can also limit radiation when objects are tested for susceptibility. Whatever be the application, the principles discussed in this book apply in detail.

Electrical conduit should not be routed along the exterior walls of a screen room. Facility power wiring, distribution panels, fans, and motors should not be located near the room. Care must be taken that *equipment ground* currents do not flow in the walls, floor, or ceiling of the room. The approach that is used is to limit all metal and electrical connections to the room to one area. This area is where the *equipment ground* and power enters the room. Any ventilation duct must be insulated from the wall of the screen room and the air path should be through a honeycomb filter that is properly bonded to the wall. The floor must also be insulated from building steel. Communications into the room can be on shielded cable provided the opening does not provide an aperture. If optical fiber is used, the support wire should not be brought into the room through the aperture. If a hole is made in the wall, it can be made into a waveguide beyond cutoff by extending the depth of the aperture. No conductor can pass through this hole without violating the shielding integrity of the room.

Users of the screen room should be aware that the fields near radiating hardware are often near fields. When these fields have a low wave impedance, they are more apt to penetrate the walls of the room. It is recommended that radiating sources be kept away from walls to limit field penetration.

N.B.

Turn off all cell phones.

The Decibel

There are many parameters in engineering that extend in value over many decades. As an example, useful voltages extend from well below a microvolt to megavolts. That is over 12 orders of magnitude. Often it is convenient to consider a logarithmic scale rather than a linear scale to discuss parameters. In the early days of telephony, engineers needed a logarithmic scale for discussing sound levels. Noise levels, for example, were millivolts while signal levels were volts. The power ratio here is a million to one. It was convenient to use a logarithmic scale, where a just discernable change in loudness was one unit.[1] This unit had to work for noise as well as voice signals. It turned out that the logarithm of the ratio of power from two signals when multiplied by 10 was just such a scale. It worked at all signal levels. The unit was called the decibel to honor Mr Bell, the telephone inventor. The definition of the decibel is

$$1\,dB = 10 \log_{10}\left(\frac{P_1}{P_2}\right) \tag{A.1}$$

where P_1 and P_2 are power levels. If the signal levels are measured as voltage on one resistor, then the ratio can be written as

$$1\,dB = 10 \log_{10}\frac{(V_1)^2 R}{(V_2)^2 R} = 20 \log_{10}\left(\frac{V_1}{V_2}\right). \tag{A.2}$$

The decibel is abbreviated as dB and in conversation it is pronounced "deebee." For those that use this measure on a regular basis, they know that 6 dB is a factor of 2, 20 dB is a factor of 10, that negative decibel

---

[1] The logarithm of a number to the base 10 is the exponent of 10 that equals that number. For example, the logarithm of 100 is 2 because $10^2 = 100$. The logarithm of 2 is 0.30103 because $10^{0.30103} = 2$. Note that $20 \log_{10} 2$ is usually rounded off to 6 dB.

Grounding and Shielding: Circuits and Interference, Sixth Edition. Ralph Morrison.
© 2016 John Wiley & Sons, Inc. Published 2016 by John Wiley & Sons, Inc.

figures imply division. For example, –6 dB means a factor of 0.5 or divide by 2 and –20 dB means a factor of 0.1 or divide by 10. The factor 50 is 100/2 and in decibel language this is 40 dB minus 6 dB or 34 dB. It takes usage to recognize this language.

In field measurement, the units can be volts, current, the H or E field, and watts. In fact, the decibel scale can be applied to nonelectrical parameters such as meters. This means that units must usually be associated with a decibel statement. It is a standard practice to refer to 20 dB V, which means 10 V. A total of 6 dB V means 2 V. It is important to know that 0 dB means a factor of 1. There is no decibel representation of 0 V as the logarithm of zero is minus infinity.

The parameters P_2 or V_2 are called reference parameters. They might be units of watts or volts, milliwatts, or millivolts as well as megawatts or megavolts. If the reference parameter is a millivolt, then 20 dB mV means 10 mV. The reference parameter must be stated or the decibel statement has no meaning. There are a group of abbreviations that have become standard. Again it takes usage to become familiar with this language. Here are a few standard abbreviations.

dB V	dB volts	The reference is 1 V
dB mV	dB millivolts	The reference is 1 mV
dB mW	dB milliwatts	The reference is 1 mW
dB μV	dB microvolts	The reference is 1 μV
dB V/MHz	dB volts per megahertz	The reference is 1 V/MHz
dBrn	dB reference noise	Used in telephony. The noise reference level is –90 dB W

The term decibel usually implies a power ratio. When the units are volts or amperes, Equation A.2 is used. When the units are power, Equation A.1 is used. When the units are a parameter such as ohms or meters, it is obvious that power does not apply. In these cases, the term decibel represents a logarithmic scale of $20 \log_{10} A/B$. Many other disciplines that need logarithmic scales have adapted the decibel for their use. An example is a specification for street paving. Roughness is described in terms of decibel inches. The reference parameter is I inch and this has little to do with power. Equation A.2 is used.

FURTHER READING

B.R. Archambeault, *PCB Design for Real World EMI Control*, Kluwer Academic Publishers, Norwell, Massachusetts, 2002.

J.R. Barnes, *Electronic System Design and Noise Control Techniques*, Prentice Hall, Hoboken, New Jersey, 1987.

S.H. Hall, G.W. Hall, and J.A. McCall, *High-Speed Digital System Design*, Wiley, Hoboken, New Jersey, 2000.

C. Harper, *High Performance Printed Circuit Boards*, McGraw-Hill, New York, New York, 2000.

H. Johnson and M. Graham, *High-Speed Digital Design*, Prentice Hall, Upper Saddle River, New Jersey, 1993.

L. Martens, *High-Frequency Characterization of Electronic Packaging*, Kluwer Academic Publishers, Norwell, Massachusetts, 1998.

R. Morrison, *The Fields of Electronics*, Wiley, Hoboken, New Jersey, 2002.

R. Morrison, *Digital Circuit Boards: Mach 1 GHz*, Wiley, Hoboken, New Jersey, 2012.

R. Morrison and Warren Lewis, *Grounding and Shielding in Facilities*, Wiley, Hoboken, New Jersey, 1990.

W.O. Henry *Noise Reduction Techniques in Electronic Systems*, 2nd Ed., Wiley, Hoboken, New Jersey, 1988.

C.S. Douglas *High-Frequency Measurements and Noise in Electronic Circuits*, Wiley, Hoboken, New Jersey, 1993.

Grounding and Shielding: Circuits and Interference, Sixth Edition. Ralph Morrison.
© 2016 John Wiley & Sons, Inc. Published 2016 by John Wiley & Sons, Inc.

Grounding and Shielding: Circuits and Interference, Sixth Edition. Ralph Morrison.
© 2016 John Wiley & Sons, Inc. Published 2016 by John Wiley & Sons, Inc.

Printed and bound by CPI Group (UK) Ltd, Croydon, CR0 4YY

17/04/2025

14658862-0001